Truck Carrier Partner 2.0.13 Tool:
Data Entry and Troubleshooting Guide
2013 Data Year - United States Version

Transportation and Climate Division
Office of Transportation and Air Quality
U.S. Environmental Protection Agency

 United States
Environmental Protection
Agency

Office of Transportation and Air Quality
EPA-420-B-14-002
January 2014

Table of Contents

Introduction

This guide is the second of two guides available to help your company participate fully in the SmartWay Transport Partnership as a Truck Carrier Partner.

The first guide, the Truck Carrier Tool "Quick Start Guide," provides a basic walk-through of the process of identifying, gathering, preparing, and submitting your data using the SmartWay Truck Carrier Tool. (From this point on, this guide will refer to the SmartWay Truck Carrier Tool as the "Truck Carrier Tool" or simply, the "Tool," for brevity.)

The Quick Start Guide may be all you need to successfully complete the Truck Carrier Tool.

This guide, the Truck Carrier Tool "Data Entry and Troubleshooting Guide," is intended to supplement the Quick Start Guide and provide more detailed information for completing your Tool submission.

The Table of Contents for this guide is designed so that you can find the relevant information on specific topics without reading the entire guide.

In this guide, you will learn how to enter the fleet composition and activity data you collected (as mentioned Quick Start Guide and the online data entry forms it references) into the Truck Carrier Tool. This guide covers:

- Downloading and setting up the Tool to run properly
- Basic organization of the Truck Carrier Tool
- Entering your data
- Viewing fleet performance summary data
- Submitting data to SmartWay
- Troubleshooting the Tool
- Appendices

WARNING!

*Completing this Tool requires a considerable amount of information about your fleet(s). There are many sections and screens to complete for each fleet. While you will have the ability to save your Tool along the way and return to it at anytime, we **STRONGLY ENCOURAGE** you to review the Quick Start Guide to understand key information about joining the partnership and preparing the required data **BEFORE** attempting to complete the Tool.*

PART 1: DOWNLOADING AND SETTING UP THE TOOL

Downloading and Setting up the Tool

SOFTWARE AND HARDWARE REQUIREMENTS

The Truck Carrier Tool was designed in Microsoft Excel. Microsoft Excel is an electronic spreadsheet program used for storing and manipulating data. Microsoft Excel Forms were used to enable the functional capabilities of the Tool.

Completing the Truck Carrier Tool requires the following software and hardware:

- A 2003 or later version of Microsoft Excel
- Excel security level set at Medium or lower
- A PC running Windows XP or newer operating system, or a Mac that is running the Windows XP operating system The tool does not currently work using the Mac operating system
- At least 10 megabytes of available disk space (more disk space may be required based on the number of companies you define in your Tool).
- Adequate memory (RAM) to run Microsoft Office
- A monitor resolution of at least 1,024 x 768[1]

Check with the user guides for your computer, online support, or your company's IT department to make sure your system is set up to use the Truck Carrier Tool.

We encourage you to make sure that you virus software is up to date, and scan your PC before putting data in the Truck Carrier Tool.

DOWNLOADING THE SMARTWAY TRUCK CARRIER TOOL

To download the Truck Carrier Tool, visit: http://epa.gov/smartway/partnership/trucks.htm

Save the Tool in a folder on your hard drive; this folder will house copies of your data and future updates.

HOW TO SET SECURITY LEVELS FOR THE SMARTWAY TOOLS

The following instructions should appear on your screen *if* you need to change your security settings before running the Tool. The instructions differ depending upon what version of Excel you use (Excel 2003, 2007 or 2010).

---

[1] The tool will also work at 800 x 600 resolution, but many of the screens will appear with scroll bars.

SECURITY SETTINGS FOR EXCEL 2003 USERS

To use the Truck Carrier Tool in Microsoft Excel 2003, you will need to have your security levels set to "Medium."

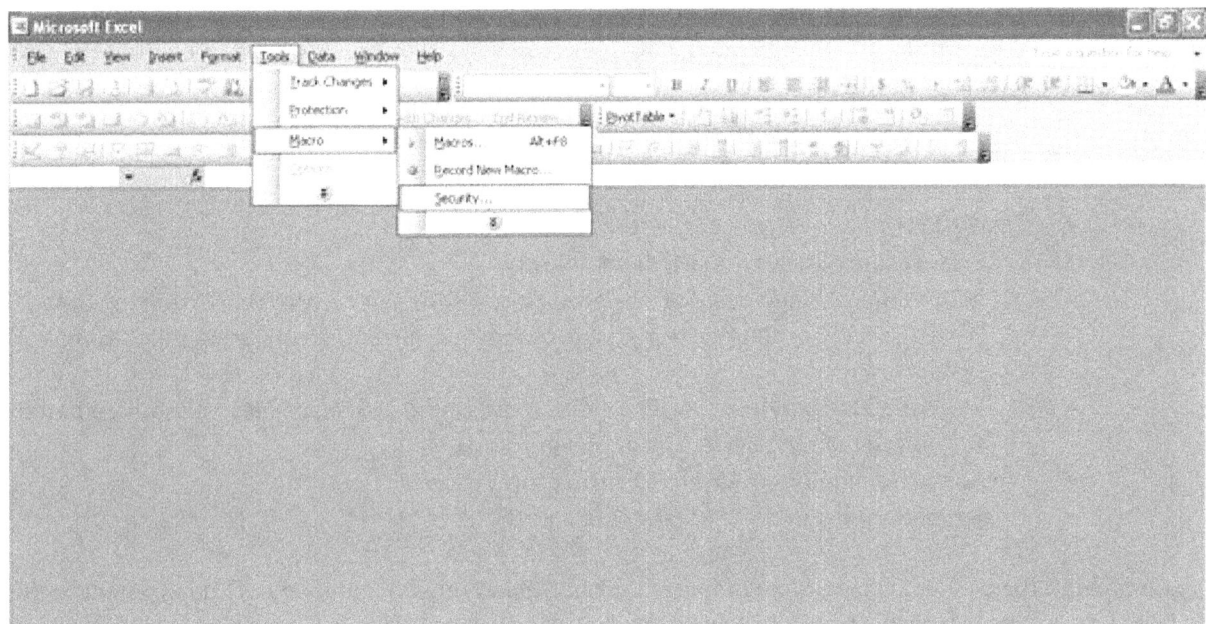

Figure 1: Macro Setting Tabs in Excel 2003

Setting Security Levels to Medium in Excel 2003

1. When using Excel 2003, on the menu bar, go to *Tools → Macro →Security Level.*

2. When the "Security" window opens, select the "Medium" level, and select **OK**.

Figure 2: Security Level Setting Screen in Excel 2003

Running the Tool in Microsoft Excel 2003

1. Save the Tool to your computer in a folder on your hard drive.

2. Go to that folder and double-click on the file to open the Tool.

3. You will see a security-warning box appear (**Figure 3**). Select the [Enable Macros] button in the security-warning box.

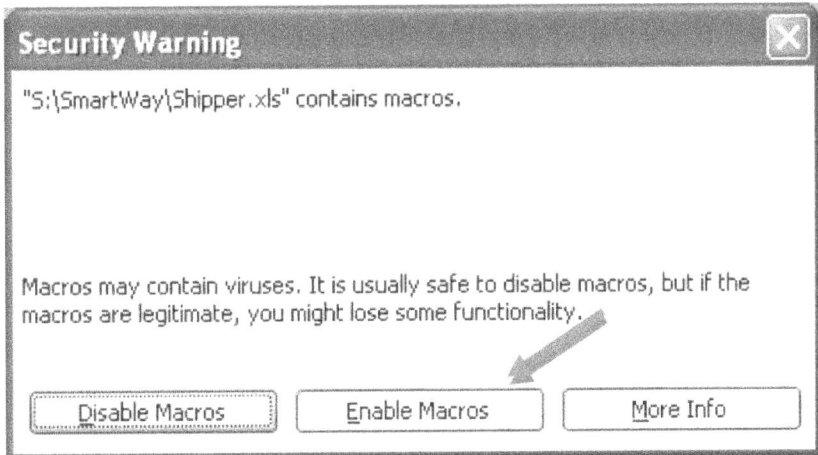

Figure 3: Screen showing "Enable Macros" button

The Welcome Screen for the Truck Carrier Tool should then appear and you will be ready to begin working on your tool.

SECURITY SETTINGS FOR EXCEL 2007 USERS

The default settings for Excel 2007 should enable you to run the Tool without any changes.

Running the Tool in Microsoft Excel 2007

1) Save the Tool to your computer.

2) Open the file, and select the [Options...] button that appears after the Security Warning just below the menu bar (**Figure 4**). Detailed instructions are also provided on the screen itself.

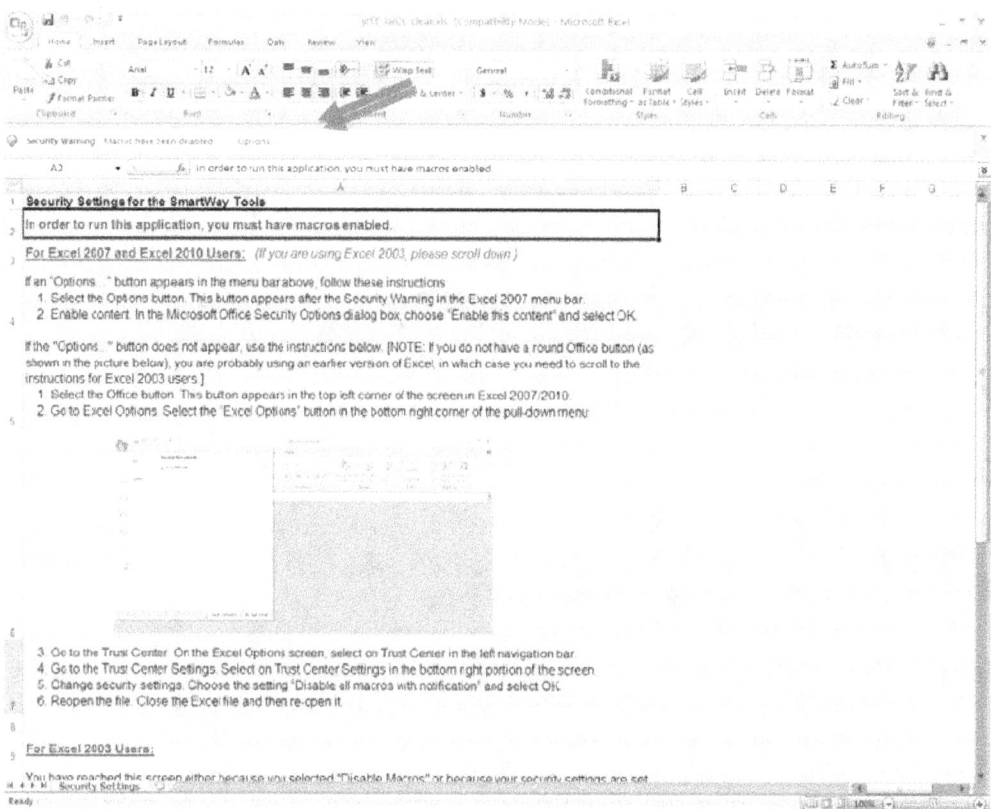

Figure 4: Security Warning Screen

3)   In the **Microsoft Office Security Options** dialog box (**Figure 5**), choose "Enable this content" and select **OK**.

Figure 5: Security Options Dialogue Box

The Welcome Screen for the Truck Carrier Tool should then appear and you will be ready to begin working on your Tool.

Troubleshooting the Security Settings in Microsoft Excel 2007

If you reach this point and the Tool does NOT open, you may have your security set too high.

To adjust your security settings, select the [button image] button (in the top left corner of the screen) and then select the [Excel Options] button in the bottom right corner of the pull-down menu (**Figure 6A**).

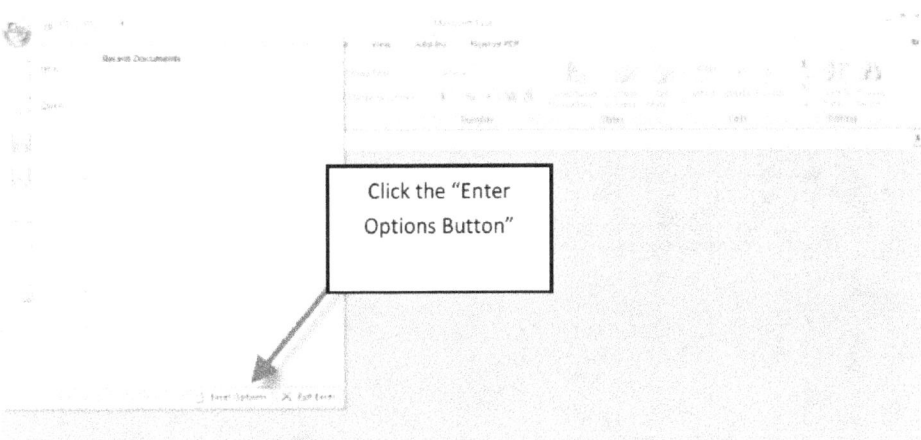

Figure 6A: Excel Options Drop-Down Menu

On the Excel Options screen, select **Trust Center** in the left navigation bar (**Figure 6B**):

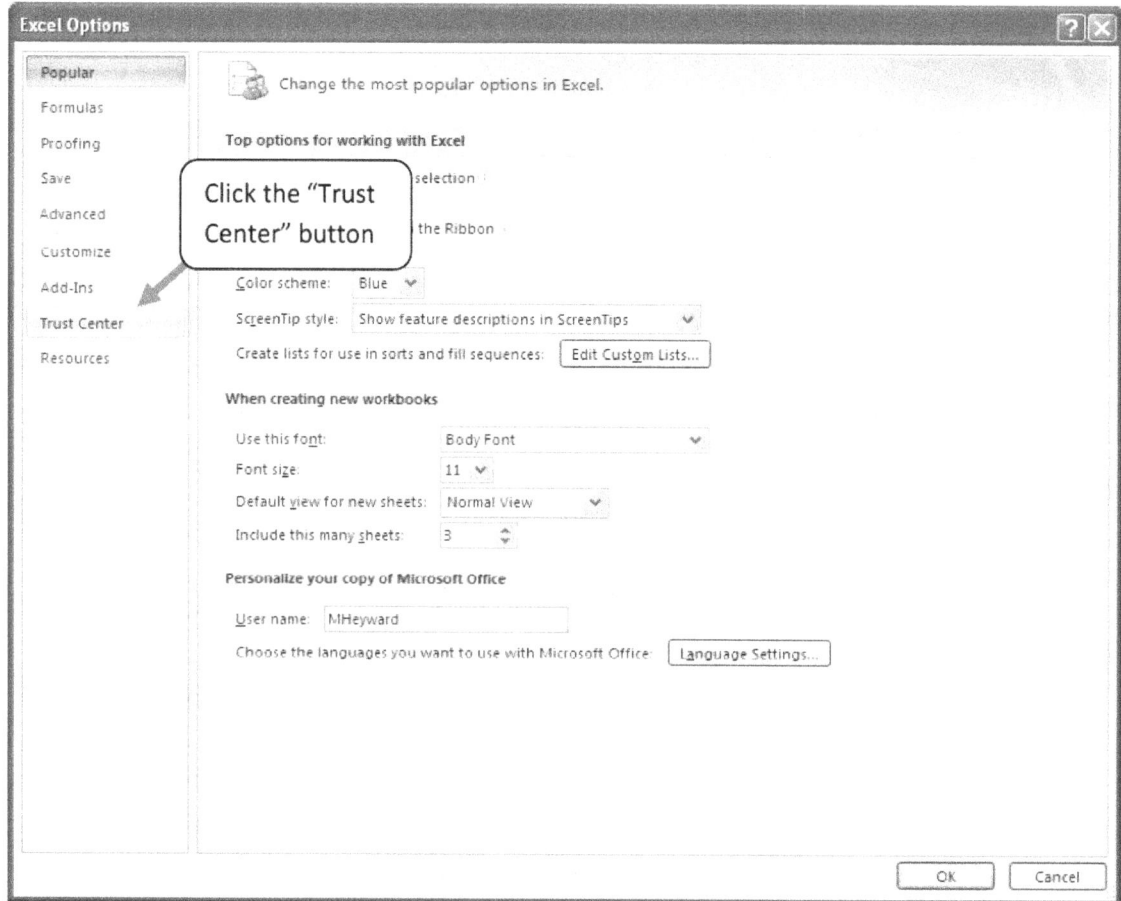

Figure 6B: Excel Options Drop-Down Menu

When the Trust Center options display opens, select **Trust Center Settings** in the bottom right portion of the screen (**Figure 7**):

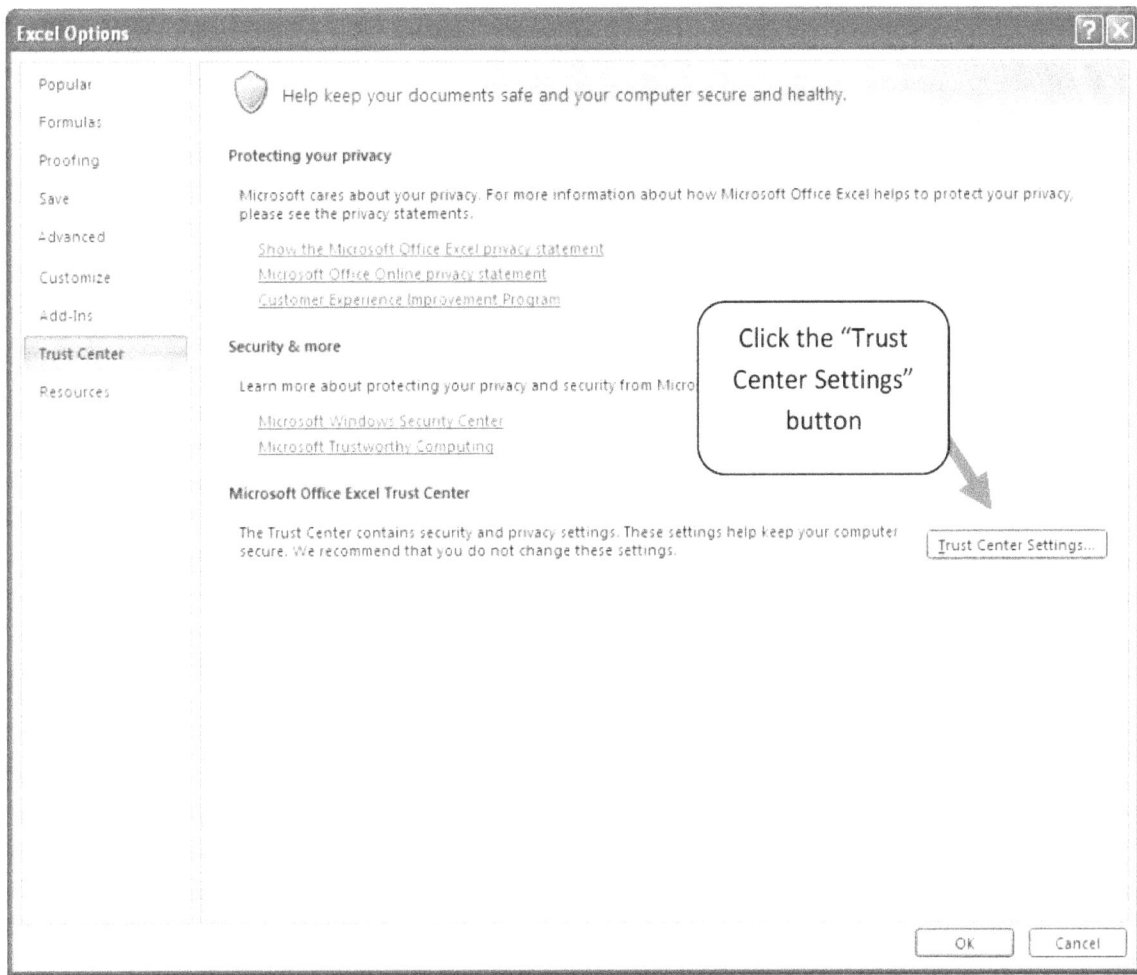

Figure 7: Trust Center Settings Screen

Choose the setting "Disable all macros with notification" (**Figure 8**) and select **OK**.

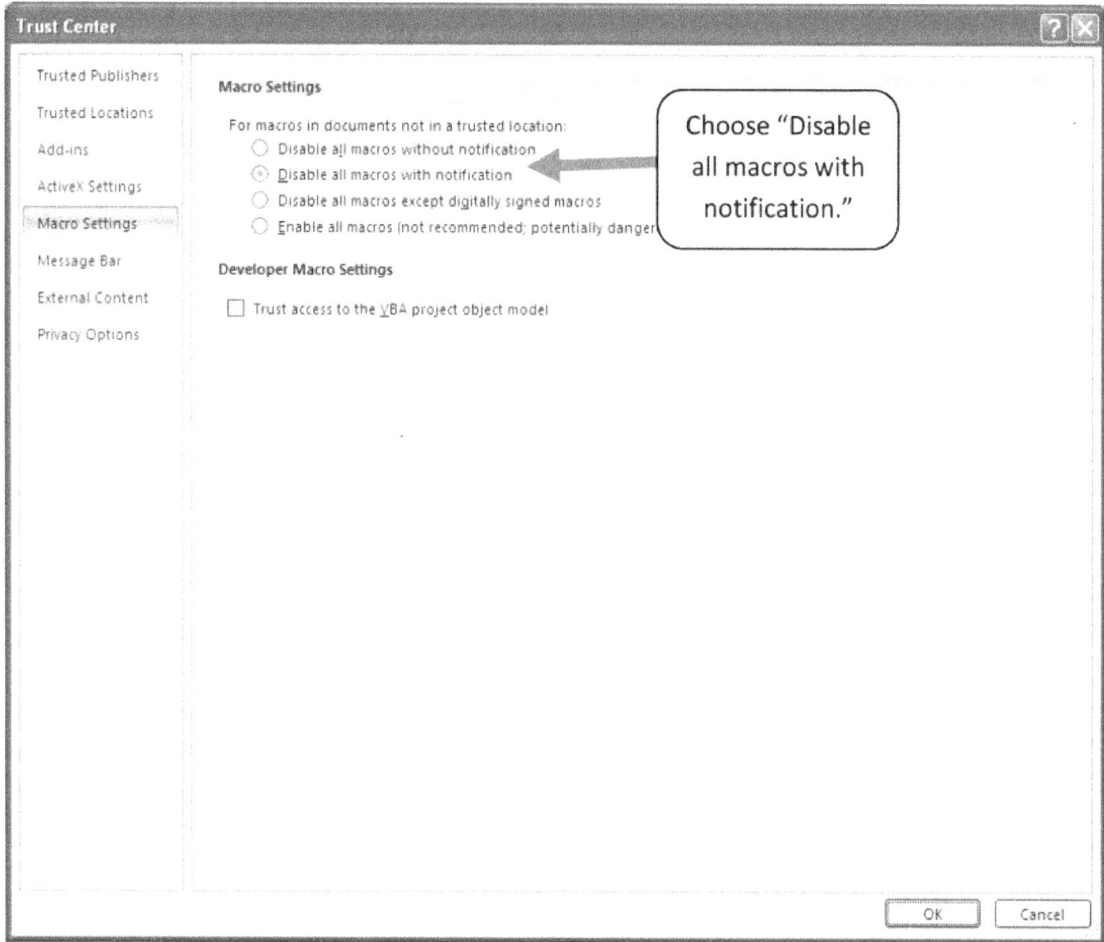

Figure 8: Macro Settings Screen

Then run the Tool.

If, at this point, the Tool does not open, review the "Software and Hardware Requirements" on page 7. If you after reviewing this section, you cannot determine how to correct the problem, contact your SmartWay Partner Account Manager.

SECURITY SETTINGS FOR EXCEL 2010 USERS

The default settings for Excel 2010 should enable you to run the tool without any changes. To run the tool:

1) Save the Tool to your computer.

2) Open the file. Depending on your Office settings, you may receive an "Enable Editing" popup. If you do, simply select the   Enable Editing   button. This will allow you to enter data into the Tool. You may only receive this popup the very first time you open the Tool.

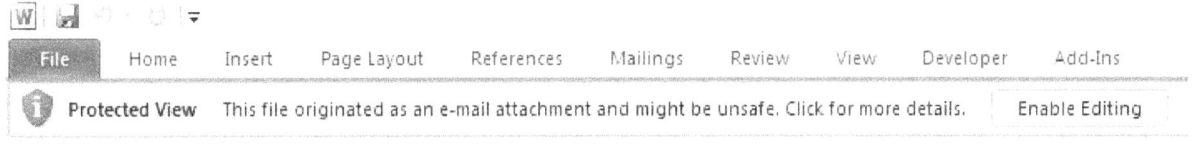

Figure 9: The Enable Editing Button

3) Depending on your Excel macro security settings, you may receive a   Enable Content   popup. If you do, simply select the   Enable Content   button. This will enable macros in the tool you just opened.

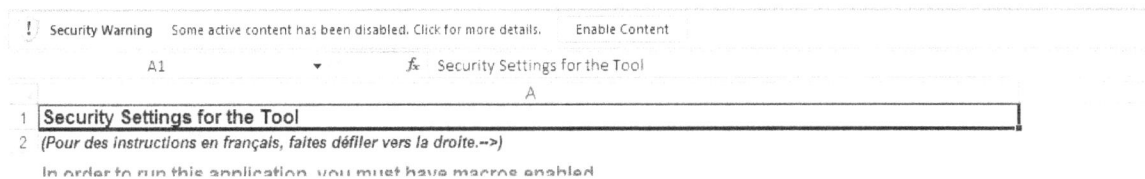

Figure 10: Security Warning Popup

If, at this point, the Tool does not open, review the "Software and Hardware Requirements" on page 7.

Otherwise, you may have your security set too high. To adjust your security settings, select the ⬤ button (in the top left corner of the screen) and then select the [🗇 Excel Options] button in the bottom right corner of the pull-down menu (**Figure 11**):

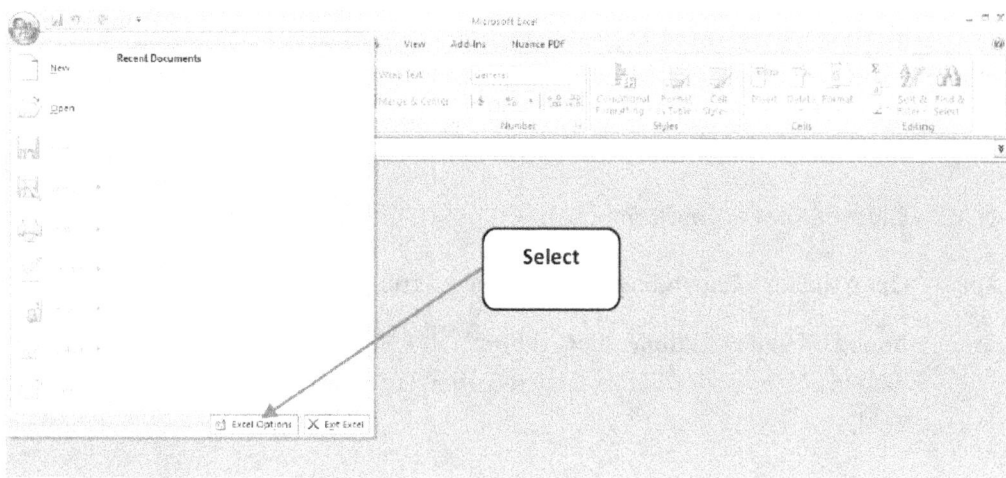

Figure 11: Excel 2010 Options Drop-Down Menu

On the Excel Options screen, select **Trust Center** in the left navigation bar (**Figure 12**):

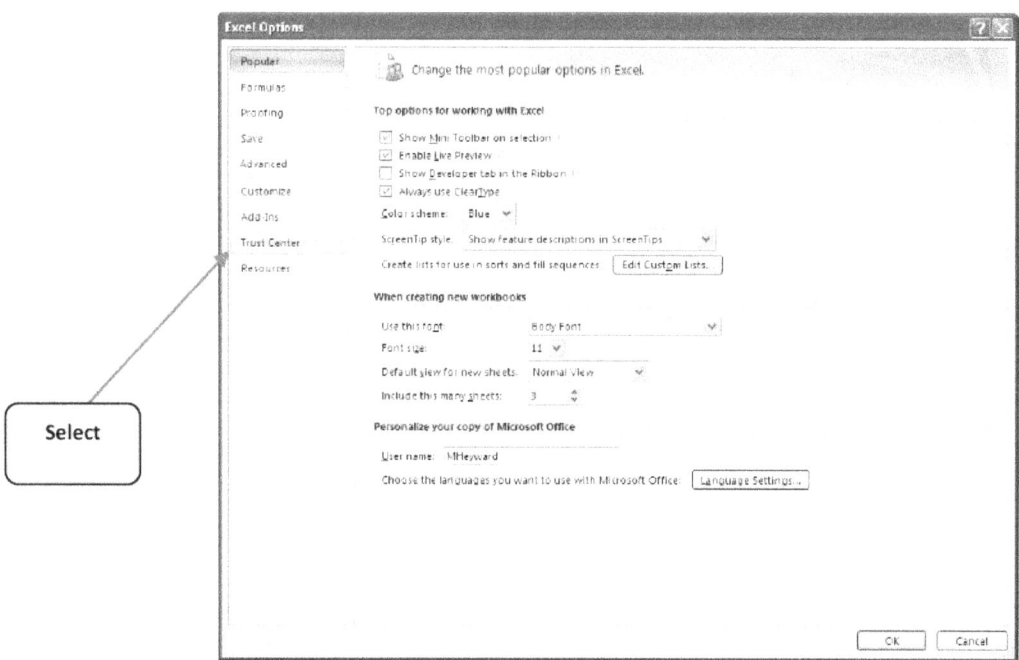

Figure 12: Excel 2010 Options Drop-Down Menu

When the Trust Center options display opens, select **Trust Center Settings** in the bottom right portion of the screen (**Figure13**):

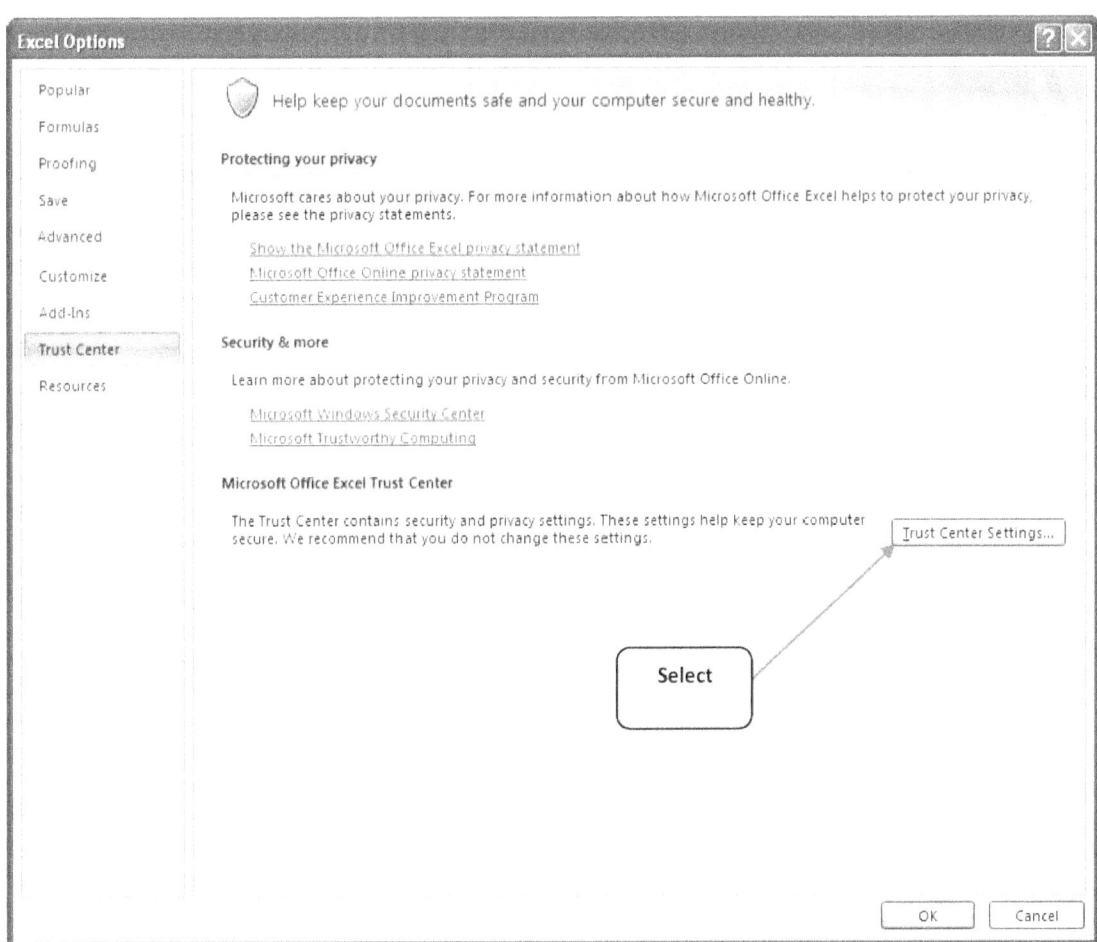

Figure 13: Trust Center Settings Screen

Choose the setting "Disable all macros with notification" (**Figure 14**) and select **OK**.

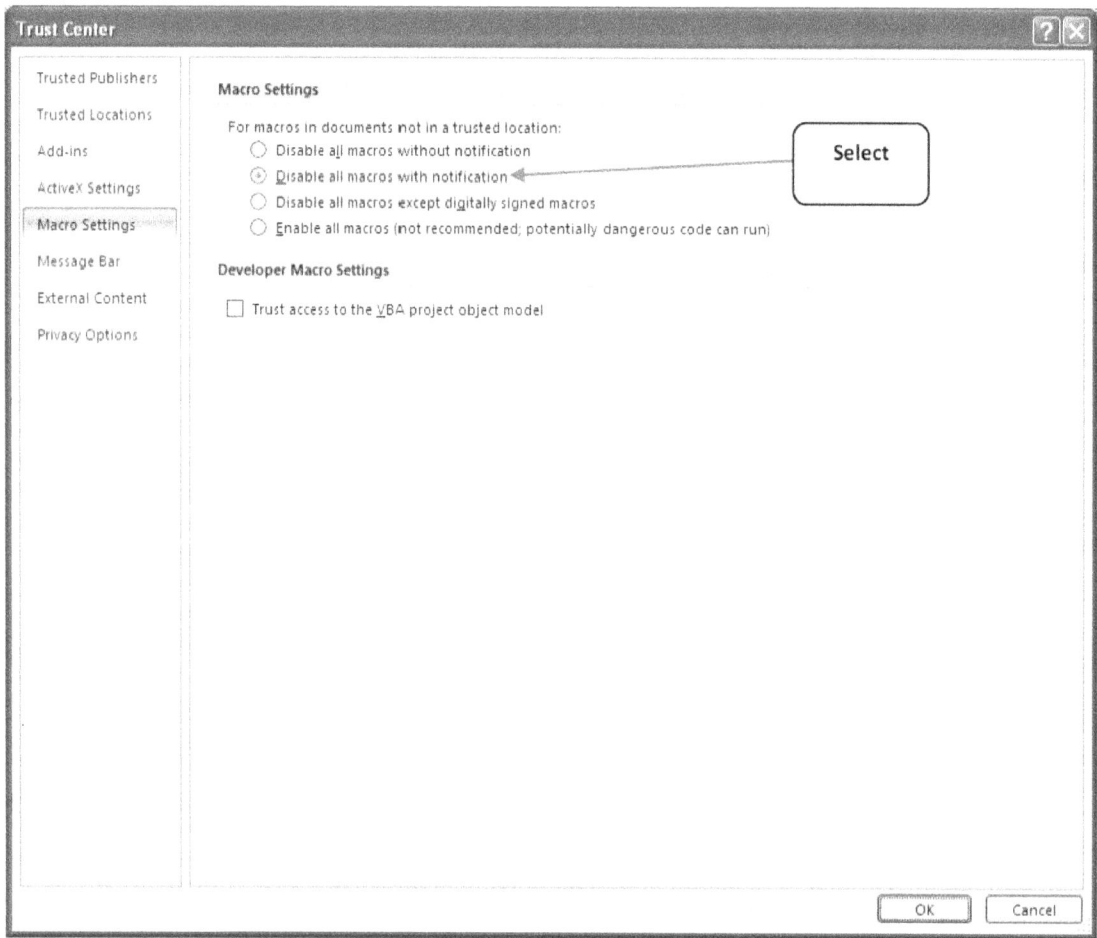

Figure 14 Macro Settings Screen

PART 2:
TOOL ORGANIZATION

Basic Organization of the SmartWay Truck Carrier Tool

The Truck Carrier Tool is the basis of the SmartWay Partnership for Truck Carriers. Completion and submission of a Truck Carrier Tool is the first step to becoming a SmartWay Truck Carrier Partner. Your Tool submission must be approved by EPA before you are officially a Partner.

The Truck Carrier Tool is organized around

- information screens
- forms or worksheets
- reports and summaries

Each screen or form opens up within a Microsoft Excel spreadsheet. The screens generally look like the one shown here.

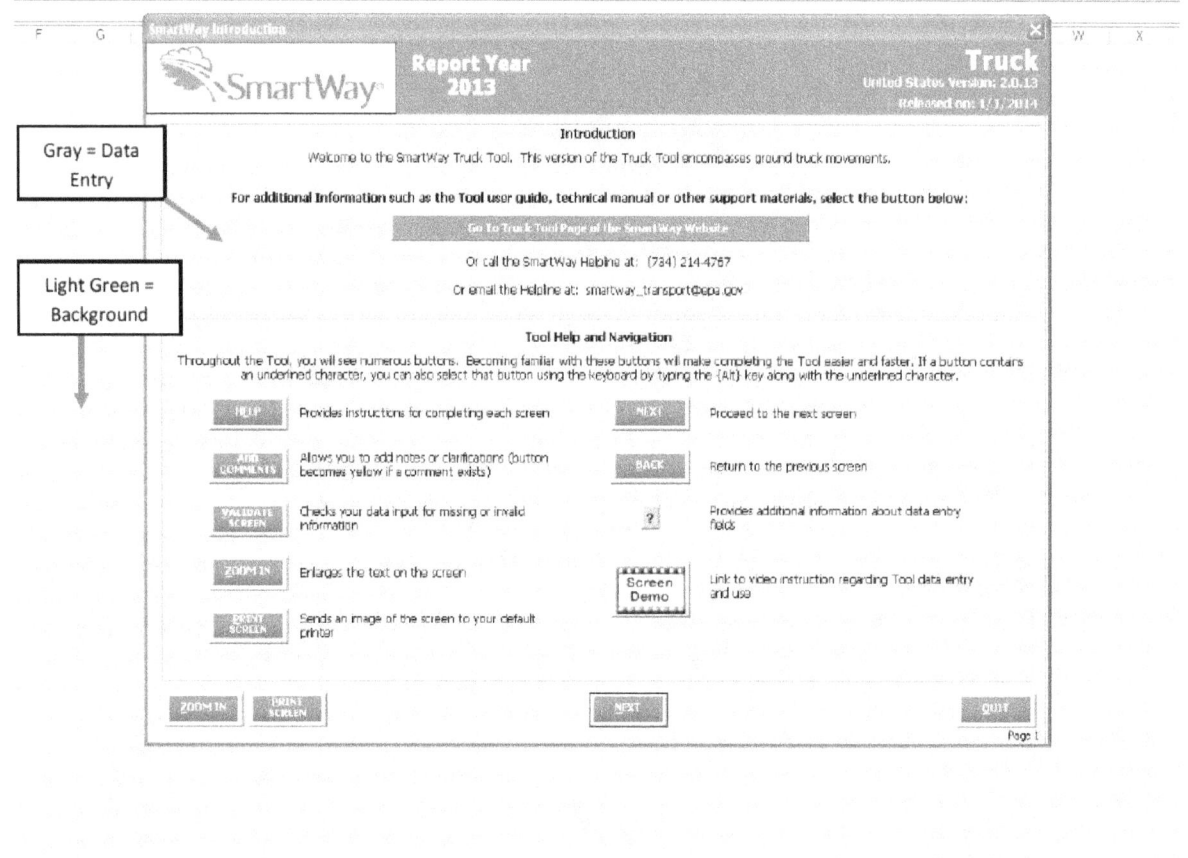

Figure 15: Screenshot of the Orientation of the Truck Carrier Tool in Microsoft Excel

The gray screen is where you enter data. The light green background is the Excel workbook; this area remains in the background is NOT used for data entry. For the purposes of your Tool submission, disregard the background workbook.

The name of each form appears at the top left-hand corner of the screen, in white text on the blue window bar, and the reporting year for the Tool is prominently displayed at the top of the screen.

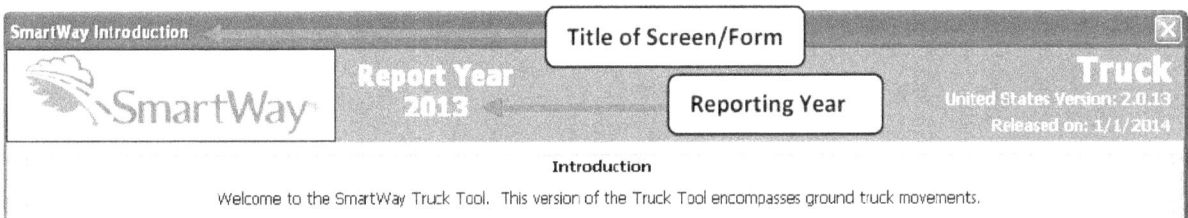

Figure 16: Screenshot of Title of Screen/Form

Each screen contains text instructions or information about the program. Where applicable, the screen will also include buttons linking to the SmartWay website or other sections of the Tool (e.g., the various data entry screens); these buttons will be shown in green and clearly labeled, as seen in **Figure 17**.

Figure 17: Screenshot of Button Link in the Tool

The screens also contain navigation buttons to direct you through the Tool.

Figure 18: Screenshot of Selected Navigation Buttons in the Tool

When new concept/topic is introduced on a screen, a small question mark ([?]) appears next to it. When you select the question mark, you will find additional definitions of terms or instructions to help you complete the screen properly. You may also find additional, detailed instructional material on how to successfully navigate and complete the different data entry screens within the Tool by selecting the

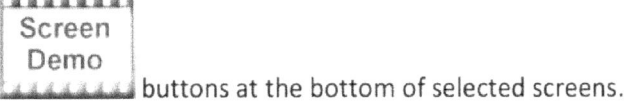 buttons at the bottom of selected screens.

Page numbers have also been added to the bottom right of all screens to facilitate navigation through the Tool and communication with your Partner Account Manager regarding any questions.

Reviewing the Introductory Screens

Before you reach your data entry section of the Tool, you will move through five introductory pages that allow you to review the basics of participation in SmartWay for truck carriers. These screens are:

- The "SmartWay Introduction" Screen
- The "SmartWay Partnership Annual Agreement" Screen
- The "SmartWay Tool Selection" Screen
- The "Required Information" Screen
- The "US/Canada Operations" Screen

The features of these screens are described below.

SMARTWAY INTRODUCTION SCREEN

The SmartWay Introduction screen is the first window that appears when the Tool is opened (**Figure 19**). This screen contains:

- a button linking to the SmartWay website where you can view and download additional information about the program, the Tool, and the technical basis for the calculations in the Tool

- descriptions of Tool Help and Navigation buttons found throughout the Tool

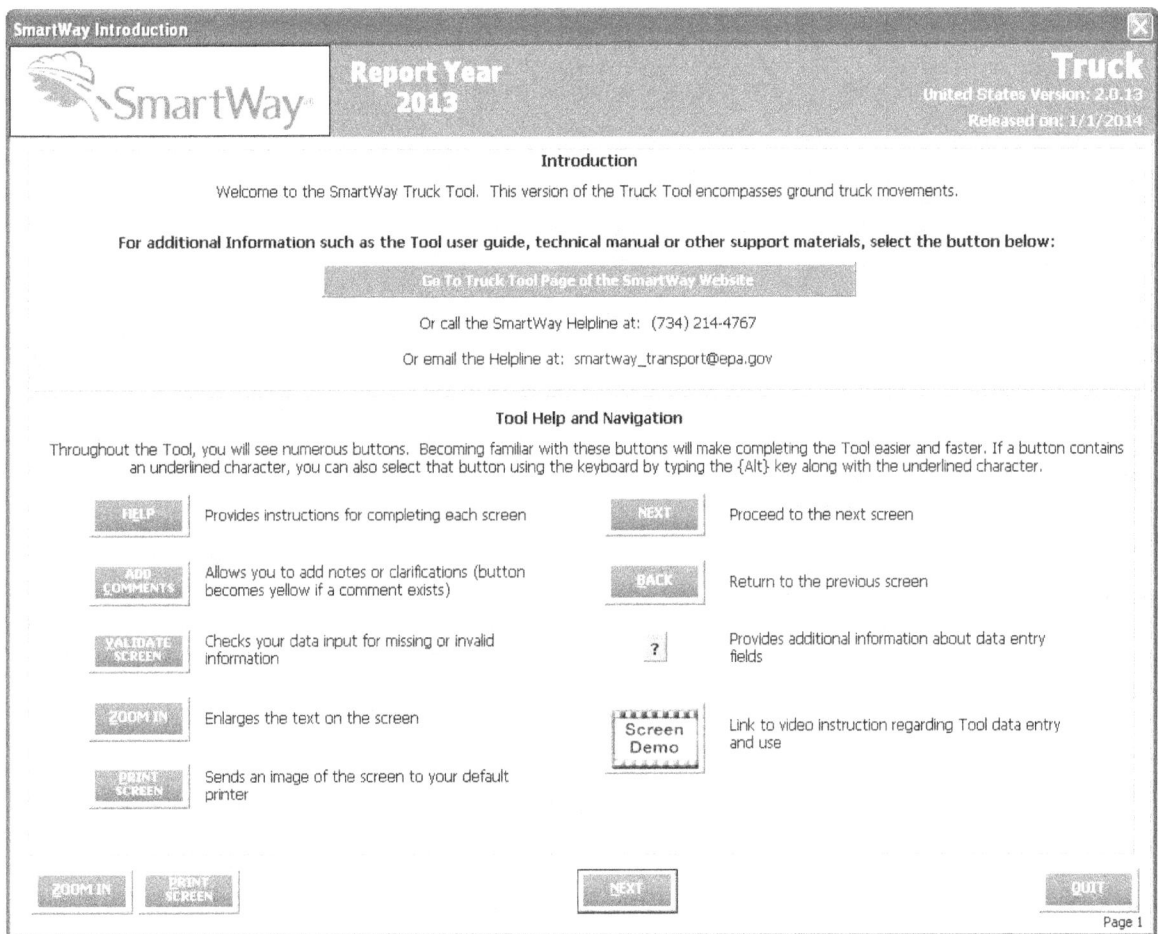

Figure 19: SmartWay Introduction Screen

Of special note is the small question mark ([?]), which appears next to new concepts when they are introduced throughout the Tool. Clicking the question mark will reveal additional definitions of terms or instructions to help you complete the screen properly.

After selecting the [   ] button on the SmartWay Introduction screen, the SmartWay Partnership Annual Agreement will appear.

SMARTWAY PARTNERSHIP ANNUAL AGREEMENT SCREEN

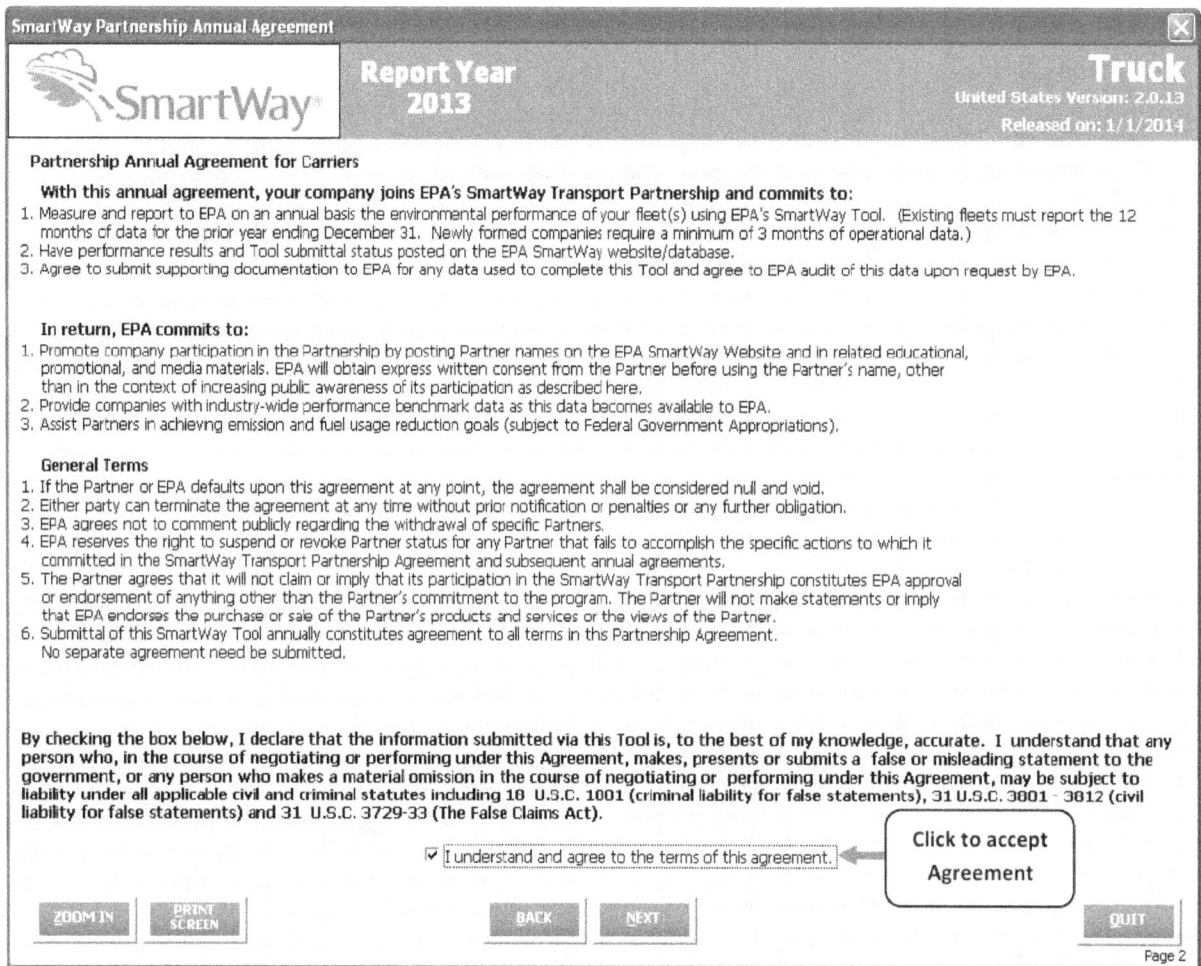

Figure 20: SmartWay Partnership Annual Agreement

Submitting a Truck Carrier Tool to EPA constitutes agreement to all terms of the Partnership Agreement, which is required for joining SmartWay and remaining a partner in good standing

All participants in SmartWay are **strongly advised to read the agreement thoroughly.**

This screen presents the agreement details; in addition, a text version of the agreement is included in the Appendix of this guide and in the Quick Start Guide for your convenience.

If you have been assigned the responsibility for completing your tool submission but are not in the position to approve the agreement, acquire any necessary approvals *before* trying to input data into the Tool.

To move to next screen, you must check the box indicating that your company accepts the terms of the agreement.

SMARTWAY TOOL SELECTION SCREEN

SmartWay offers several tools tailored for different business models and fleets. Most truck carriers will use the Truck Carrier Tool; however, it is advisable to review this screen in the Truck Carrier Tool to make sure it is the best option for your operations.

This screen provides basic information on three SmartWay tools (Truck, Logistics, and Multi-modal) appropriate for different types of trucking operations. Your choice of tool will depend primarily on your transport modes and the amount of business you contract to other companies.

Follow the flow chart in **Figure 21** to determine which tool is most appropriate for your fleet(s).

Select the [Click Here for More Information] button for further details.

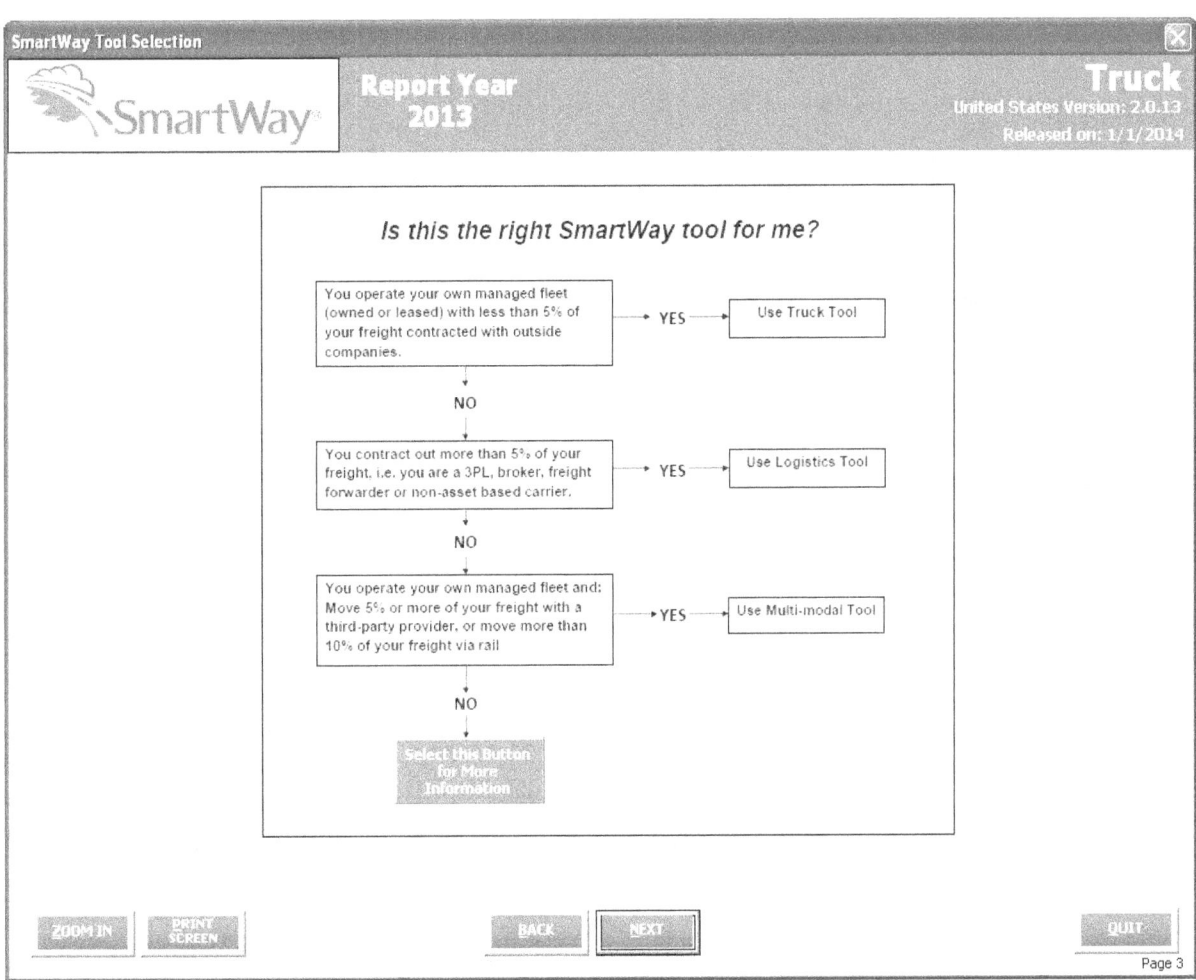

Figure 21: Determining Which Tool is Appropriate for Your Fleets

REQUIRED INFORMATION SCREEN

This next screen summarizes the information needed to complete the Truck Carrier Tool.

You may select [   ] to create a hard copy to reference as you complete the rest of the Tool.

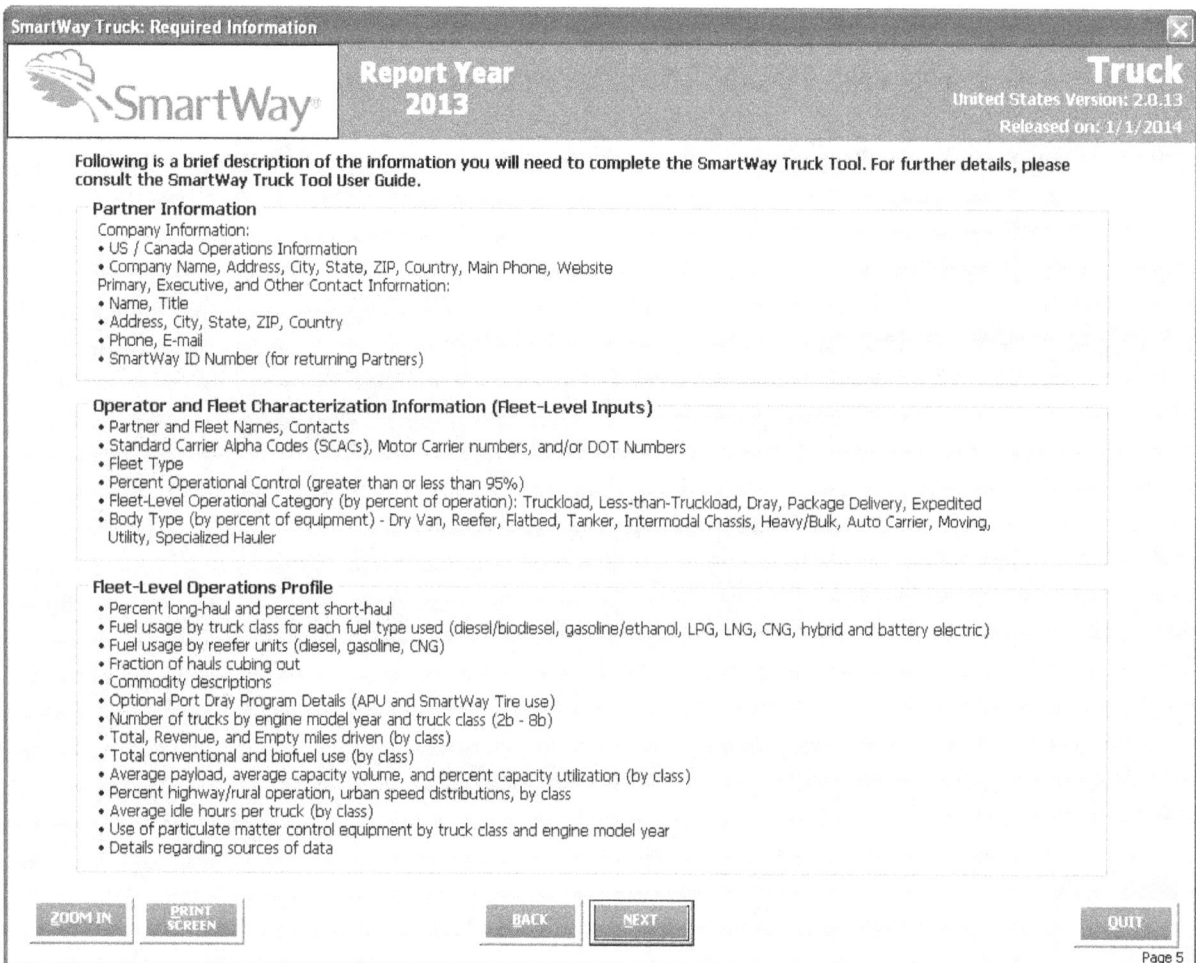

Figure 22: Summary of Required Information

While this screen serves as a reminder of your data input needs, refer to the Truck Carrier Tool Quick Start Guide for more tips and worksheets to help you gather your data for efficient and accurate Tool completion.

Once you have navigated through the four introductory screens, you will be taken to the SmartWay Truck Carrier Tool Home screen.

US/CANADIAN OPERATIONS SCREEN

The last screen asks for information about any operations you have in both the US and Canada.

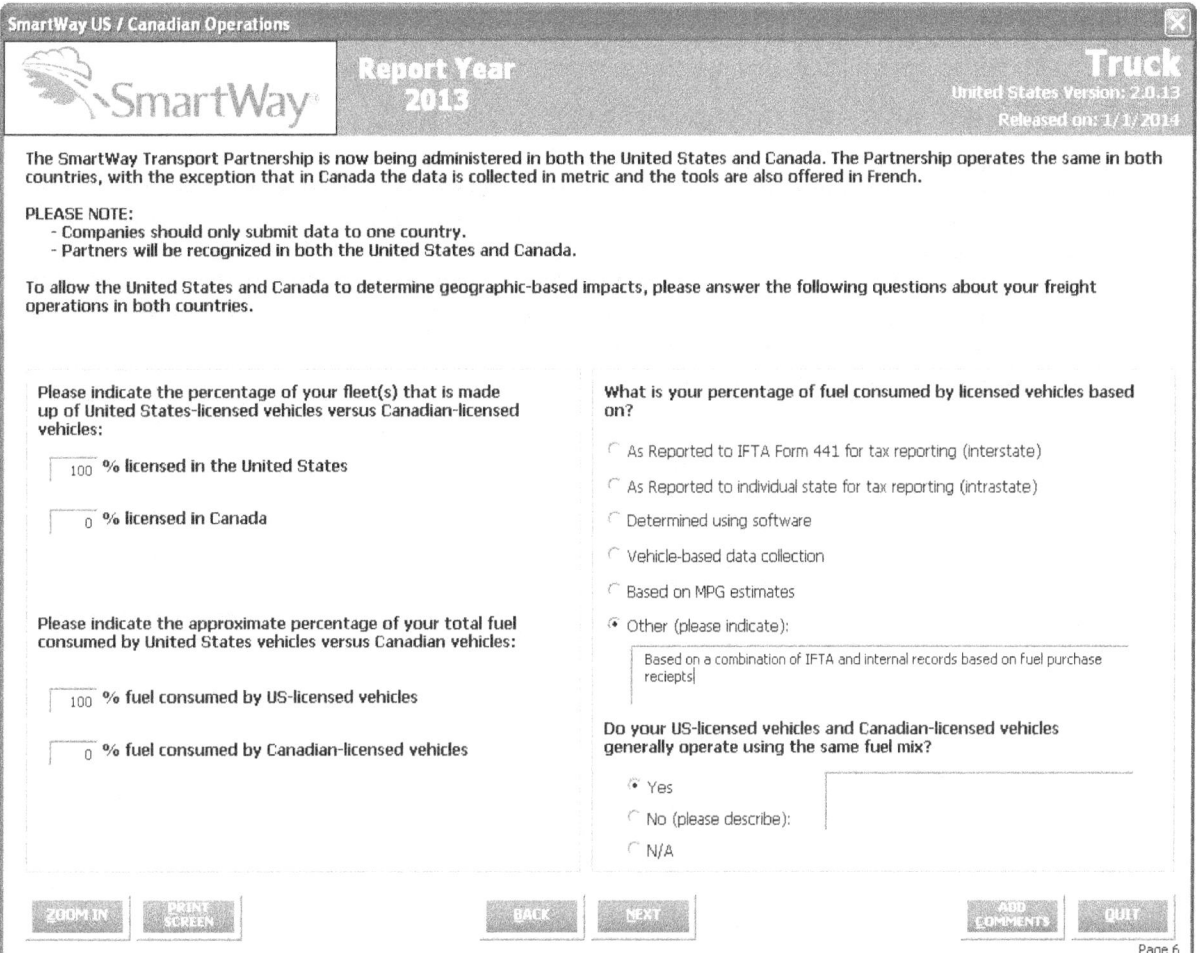

Figure 23: Screenshot of the US/Canadian Operations Screen

The SmartWay Transport Partnership is now being administered in both the United States and Canada. The Partnership operates the same way in both countries; however, data collection for Canadian Partners is collected in metric units and there are French translations of all Tool screens and guidance.

If your company operates in both the United States and Canada, note that you should ONLY SUBMIT ONE TOOL. You may select either the US Tool or the Canadian Tool, and your partnership participation will be recognized in both countries.

To allow the United States and Canada to accurately determine the impacts of freight operations in each country, you are asked to complete a single screen (**Figure 23**) indicating:

- The percentage of your fleet(s) that utilizes vehicles licensed in the United States vehicles vs. vehicles licensed in Canada.
- The approximate percentage of your total fuel consumption that is US-based vs. Canada-based
- The source of your total fuel consumption estimates
- The degree to which your Canadian fleets and US fleets use the same fuel mix

Basic Overview of How to Complete the Tool

All data entry screens are reached by starting with the Home screen.

Figure 24 displays the structure of the Home screen.

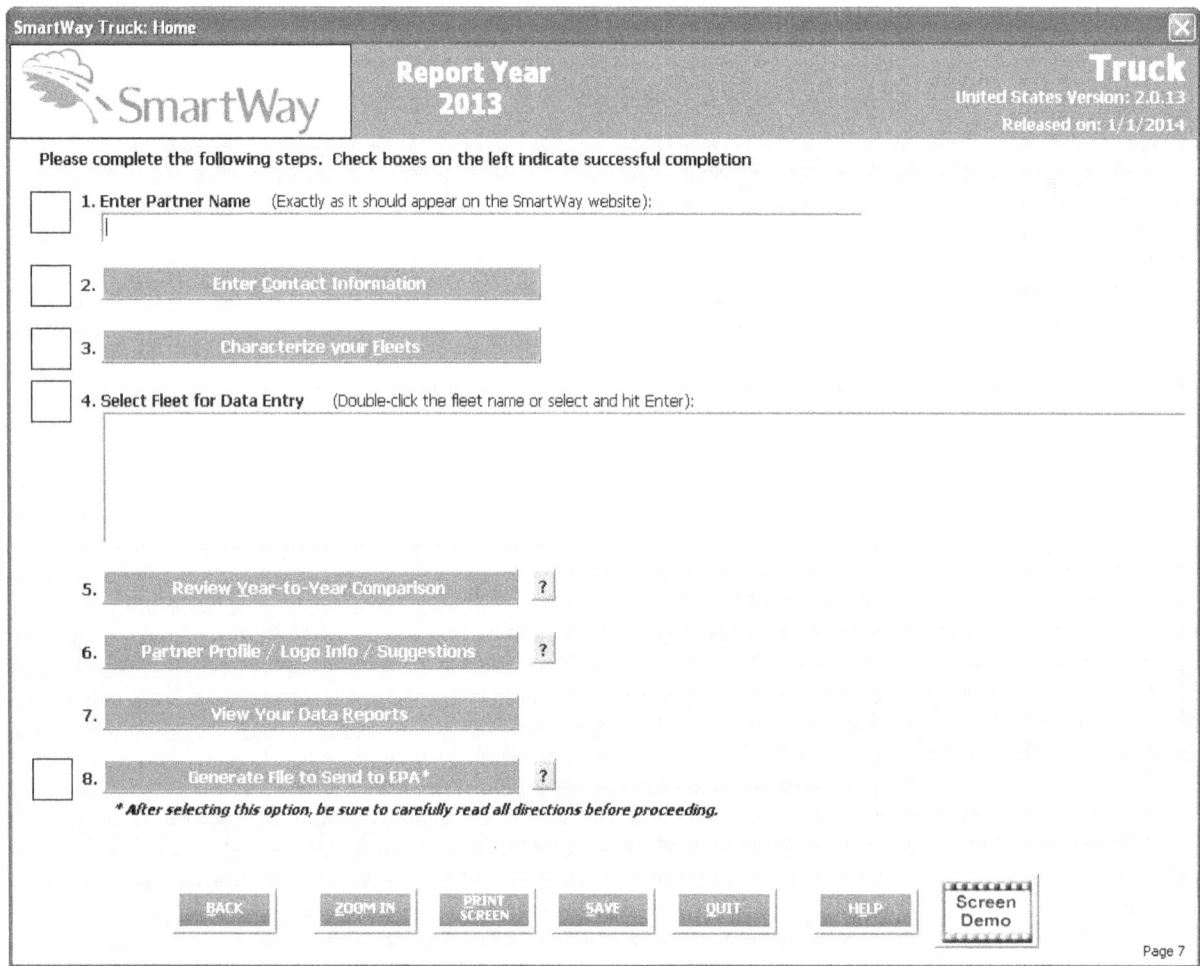

Figure 24: Main Tool Navigation or "Home" Screen

SECTIONS OF THE TOOL

The Truck Carrier Tool Home screen contains **eight sections**. Each section links to additional screens or worksheets within the Tool which are described below:

Section 1: **Enter Partner Name (data field):** Specify your company's Partner Name, exactly as you want it to appear on the SmartWay website.

Section 2: **Enter Contact Information (button):** This button takes you to a screen that asks for general company contact information, a primary SmartWay point of contact, and an executive-level contact. Additional contacts may also be included.

Section 3: **Characterize your Fleets (button):** This button takes you to a screen that asks you to define all the fleets your company operates and provide information describing their operation. Once these parameters are defined, the software will enable you to generate blank data entry forms for each fleet.

Section 4: **Select Field for Data Entry (list):** This section allows you to select the fleet for which you will be entering performance and fleet composition information necessary to calculate fleet efficiency metrics for your fleet; a list will indicate the fleet(s) defined and characterized in section 3.

Section 5: **(Optional) Review Year-to-Year Comparison of Fleets (button):** This optional step allows you to compare previous year data entries to one another or to the data entered for the current year. The comparison reports provide a useful tool for performing quality control of current data, as well as a method for evaluating trends in operation changes, activity levels and fleet performance over time.

Section 6: **(Optional) Partner Profile/Logo Info/Suggestions (button):** Here you can provide information about your company's environmental accomplishments, learn about SmartWay Logo terms and requirements, and give EPA feedback about the SmartWay program.

Section 7: **(Optional) View Your Data Reports (button):** Here you can view final summaries of your data including all data inputs, fleet performance summaries, an "out of range" report (summarizing inputs that are higher or lower than expected values) and a summary of comments you have entered.

Section 8: **Generate File to Send to EPA (button):** This button creates a version of the Tool (in XML format) for you to send as an attachment in an e-mail to your Partner Account Manager (PAM). Selecting the **OK** button on this screen does *not* automatically submit the file to EPA; you still need to submit it to EPA by attaching it in an e-mail.

ENTERING YOUR DATA

With the exception of Section 1, clicking on the buttons or list items in Sections 2 through 8 will take you to additional screens and worksheets that comprise the data entry segments of the Tool.

The first four sections of the home screen are mandatory and MUST be completed in order. These sections comprise all the data collection steps needed to complete your SmartWay Truck Carrier Tool. After they are completed, you can review your output and/or submit your Tool to EPA.

Once you complete each mandatory step, a ✓ will appear on the left of the screen.

VALIDATING YOUR DATA

The Truck Carrier Tool includes range checks and other validation rules to help identify potential data entry errors and/or unusual data values. To identify potential data problems and ensure a high quality data submission, select the [   ] button before moving on to the next screen.

The Tool will identify any potential data problems on that screen and prompt you to modify the entry or provide a text explanation for legitimate anomalies.

SAVING YOUR DATA

You can save the data you have entered at any time by selecting the [   ] button that appears at the bottom of all screens (including the Home screen). EPA recommends saving your data frequently if you are entering information for large numbers of fleets and/or vehicle classes.

REVIEWING YOUR DATA

Each screen within the Tool has a [   ] button. To generate a hard copy of screen text or your data inputs, select this button. The screen will be printed on your default printer. Alternately, you can return to the Home screen, select the [View Your Data Reports] button, identify the report of interest using the Reports Menu, and print them out for review. The data reports provided by the Truck Carrier Tool are discussed further in the [View Your Data Reports] section of this guide.

PARTNER PROFILE/LOGO INFO/SUGGESTIONS

The Truck Carrier Tool includes an optional section that allows you to provide EPA with additional information regarding your company's environmental stewardship, potential use of the SmartWay Logo, and general feedback regarding the SmartWay program.

SUBMITTING DATA TO SMARTWAY

Detailed instructions on properly submitting your data to EPA is included in this guide on page 99.

PART 3:
SECTION-BY-SECTION DATA ENTRY GUIDANCE

Preparing for Data Entry

To participate in SmartWay, truck carriers need to gather the following essential information to complete the Truck Carrier Tool:

- The official company name EXACTLY as you would like it presented on the EPA website
- Company contact information
- Contact details for your Primary Contact
- Contact details for an Executive Contact (cannot be the same as the Primary Contact)
- Split between US/Canada operations
- Quarterly IFTA statements[2] (for activity data) for the reporting calendar year
- Fleet details for all fleets you control:
    - SCACs, MCNs, or DOT number information
    - Total inventory of vehicles in your fleet(s), sorted by vehicle class and engine model year, body type, and operational category for the reporting calendar year
    - Total miles, revenue miles and empty miles
    - Total diesel, biodiesel and other fuel type use by class
    - Reefer fuel use by class (if applicable)
    - Average payload, average capacity volume, and percent capacity utilization by class
    - Average idle hours per truck
    - Use of particulate matter control equipment by truck class and engine model year (if applicable)
- Data sources for all data to be entered
- SmartWay ID number (if this is not your first Tool submission)

This data must be provided for all of your company's fleets. This data reflects the amount of freight carried by each carrier, the distance that freight is carried, and the fuel consumed to carry the freight.

The next four sections of this guide explain how to enter your data on each of the required and optional screens. You must complete these first four sections of the Home screen in order.

---

[2] If applicable – for Class 7, 8a and 8b trucks only.

Section 1 Data Entry: Enter Partner Name

SECTION 1 OVERVIEW

Section 1 of the Truck Carrier Tool asks you to "Enter Partner Name."

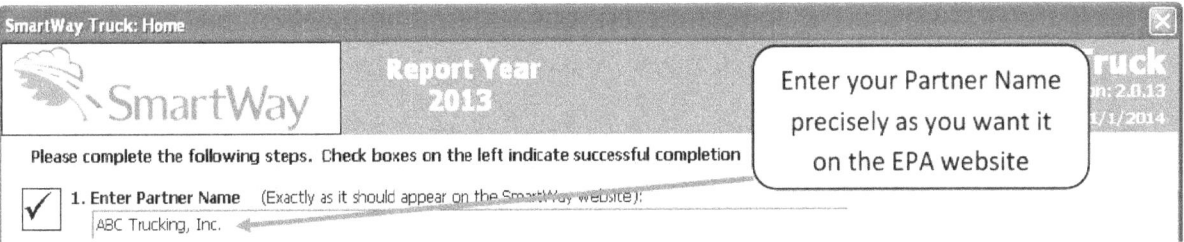

Figure 25: Field for Entry of Partner Name on Truck Carrier Tool Home Screen

EPA publishes your company's official name on the EPA SmartWay website and in the SmartWay Carrier Data File. This is how your customers will know that you are a SmartWay Partner, and how SmartWay Shippers will be able to identify your fleets for their reporting purposes. The name that EPA lists is known as your "Partner Name" and is defined within Section 1 of the Truck Carrier Tool.

Therefore, it is essential that you specify your company's Partner Name EXACTLY as you want it to appear on the SmartWay website.

Pay special attention to proper capitalization, abbreviations, and punctuation, and remember that EPA will use whatever you enter EXACTLY as reported.

Steps for Entering Partner Name

1. Type your Partner Name EXACTLY as you would like it to appear on the SmartWay website in the field as indicated.

2. Proceed to Section 2 to enter contact information.

Section 2 Data Entry: Enter Company and Contact Information

SECTION 2 OVERVIEW

The Contact Information section is where you identify all points of contact between EPA and your company that are related to your participation in SmartWay.

Section 2 asks you to click a green button labeled "Enter Contact Information."

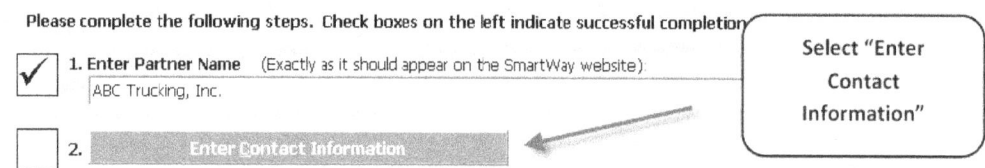

Figure 26: Select Contact Information Button on "Home" Screen

You will then be taken to the Contact Information Screen. The Partner Name entered on the **Home** screen will appear automatically at the top (see arrow in **Figure 27**).   On this screen you establish the following details about your company:

1. Partner Information includes details about the company location, phone numbers, and web address.
2. The Primary Contact is the person assigned to work directly with EPA regarding timely and accurate Tool submission, and is responsible for assembling information to complete/update fleet data; completing and updating the Tool itself; maintaining direct communication with SmartWay; and keeping interested parties within the company apprised of relevant developments with SmartWay. (NOTE: To ensure that e-mails from SmartWay/EPA are not blocked, new Primary Contacts may need to add SmartWay/EPA to their preferred list of trusted sources.)
3. The Executive Contact is the company executive who is responsible for agreeing to the requirements in the SmartWay Partnership Annual Agreement, overseeing the Primary Contact (as appropriate), and ensuring timely submission of the Tool to SmartWay. The Executive Contact also represents the company at awards/recognition events. This person should be a Vice President or higher-level representative for the company.
4. Other Contacts include any other company representatives that have a role in your participation in SmartWay. These may include representatives from other business units, media/public relations, staff, or anyone who has responsibility for specific fleets beyond the person listed as the Primary Contact. This is especially important as you will be identifying individual fleets in the Tool and will be assigning contacts to each fleet separately.

NOTE: The Primary and Executive Contact information is mandatory and these must be different people. Duplicate contacts are not allowed to ensure EPA has access to at least two people for Tool submission follow-up.

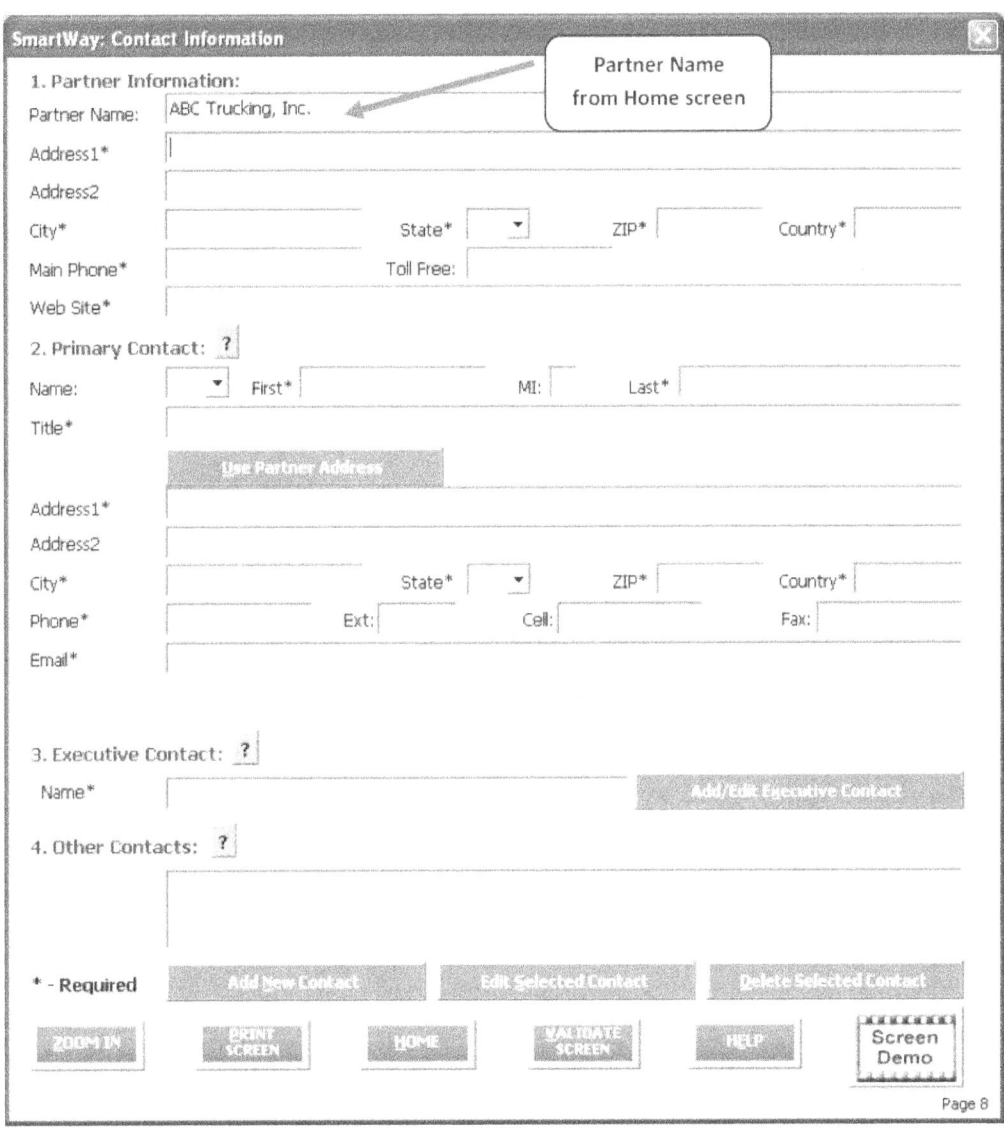

Figure 27: Blank Contact Information Screen

Each field marked with an asterisk must be filled **out.** You will not be able to submit the Tool to SmartWay without this information.

Steps for Entering Contact Information:

1. **Enter the Partner Information details** in Section 1 of the Contact Information screen.

2. **Enter Primary Contact details** in Section 2. If the Primary Contact's address is the same as the company, you can select the [Use Partner Address] button to automatically fill in the address section of this record.

3. **Enter the Executive Contact details** in Section #3 by selecting the [Add/Edit Executive Contact] button to the right; enter the required data. Note that you MUST have at least two different contacts on the Contact Information screen, and the Primary and Executive Contacts must be different.

4. **Enter Other Contacts** (if applicable) in Section 4 by selecting the [Add New Contact] button. A new contact field will appear, labeled **Other Contact Information**. Enter the first Other Contact then select **OK** when done. You can add more contacts by selecting "Add New Contact" again. If you wish to edit an existing contact's information, highlight the name you wish to edit and then select the [Edit Selected Contact] button. You can remove an existing contact by highlighting the contact and then selecting the [Delete Selected Contact] button. Duplicate contacts are not allowed.

5. Select [ ] at the bottom of the screen. If any information is missing, a dialogue box will appear informing you what additional information is required.

6. When finished, select the [ ] button to return to the Home screen and proceed to Section 3.

Section 3 Data Entry: Characterize your Fleets

The third section of the Tool is the "Characterize your Fleets" section.   This is the section where you will define the various components of your fleets.

The Truck Carrier Tool allows you to assess your operations by defining multiple fleets. If a customer has the ability to hire a specific fleet by name or type, you should add that fleet to the list of fleets that you will characterize in the Truck Carrier Tool. Do not include internal company fleet definitions or designations—only include separate fleets as they would be identified and hired by your customers.

There are four screens in the Fleet Characterization section of the Tool:

- Identify Fleets
- Fleet Details
- Operation Categories
- Body Types

Each screen, and the steps required to complete it, is described below.

"IDENTIFY FLEETS" SCREEN OVERVIEW

The Identify Fleets screen is shown below:

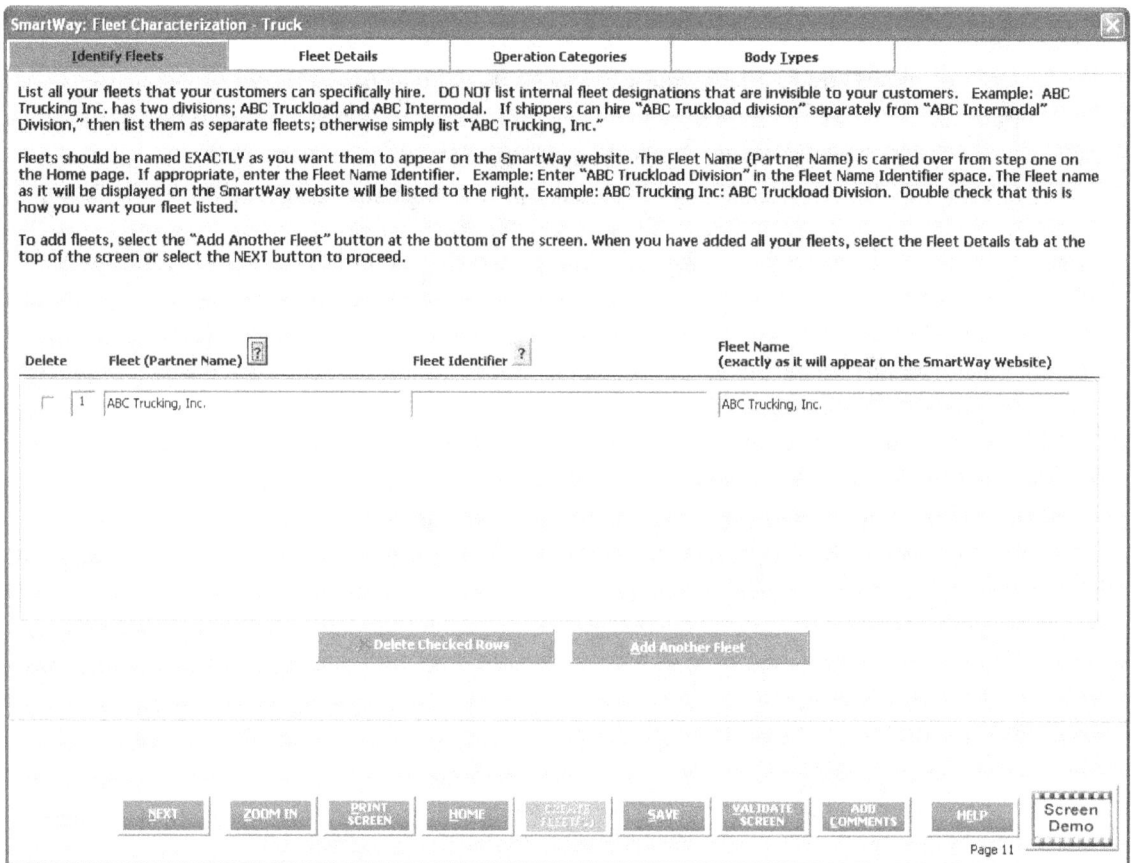

Figure 28: Identify Fleets Screen

You will be asked to establish a name for each fleet that a customer can hire. The Truck Carrier Tool combines your Partner Name (as defined on the Home screen in Section 1) with a Fleet Identifier that you provide on the Identify Fleets screen to establish your final Fleet Name.

Each of your Fleet Names will begin with the name of your company. This fleet "prefix" will be whatever you enter in the Fleet (Partner Name) field on the Identify Fleets screen. By default, this field is automatically populated with the first 50 characters of the Partner Name that was entered on the Home screen; however, you may change the prefix here by simply typing your desired prefix into the field below the heading Fleet (Partner Name). You should specify the text so that it appears EXACTLY as it you want it to show within each fleet name. Pay special attention to proper capitalization, abbreviations, punctuation, and other terms (such as Inc. or Ltd.) and adjust as necessary.

Remember that this prefix will be automatically inserted at the start of each of your Fleet Names in the SmartWay Carrier Data File. Whatever you enter for Fleet prefix for the first fleet will automatically be used for any additional fleets you add. Similarly, any edits you make to the Fleet prefix for the first fleet will automatically be reflected on each subsequent fleet.

To define each fleet name, you will be required to enter the Fleet Identifier. This should be an identifier recognizable to your customers.   Again, make sure to specify each Fleet Identifier name exactly as you want it displayed in the SmartWay Carrier Data File, including proper capitalization, any abbreviations, and punctuation. Remember that it will automatically be combined with the Fleet prefix (Partner Name) field.

Once you have added your Fleet Identifiers, the complete fleet name (with Partner Name and Fleet Identifier combined) will appear in the grayed-out box on the far right of the **Identify Fleets** screen.

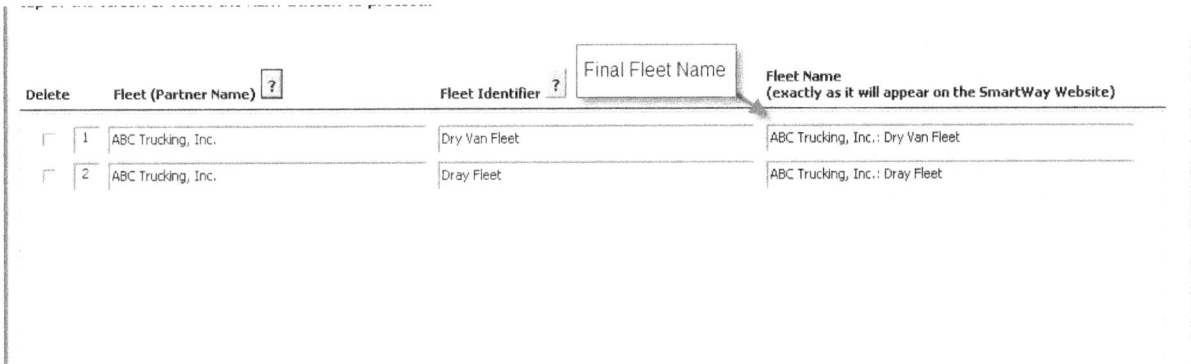

Figure 29: Example of final Fleet Name formating

NOTE: If you have only one fleet, you may leave the Fleet Identifier field blank, in which case your fleet name will simply be presented as the "Partner Name" you entered.

If you have more than one fleet, you will need to add the fleets separately. To enter another fleet, select the ⬛ **Add Another Fleet** button.

Adding Comments assists SmartWay Tool reviewers in approving your Tool as quickly as possible. Your comments help reviewers understand your business operations. Any details that you can provide related to your operations will speed up approval time.

The ⬛ button located at the bottom of the screen allows you to enter notes about the data collection process, your assumptions and methods, or other information. These details could prove useful for your reviewer when you or someone else completes the Tool next year.

The ⬛ button will be highlighted in yellow on your screen if comments have been added for a particular screen.

The button will then read ⬛ to indicate to your reviewer that there are comments to be read.

If, at a point later in the data entry process, you realize that you need to add a new fleet or delete an existing fleet, you can return to the Identify Fleets **screen.**

To add a new fleet, follow all of the instructions on the screen regarding defining your fleets, including

selecting the ▢ button at the end of the process. When you select this button, the system will create blank data entry forms only for the new fleet(s) you have added; the existing fleets will not be affected.

If you need to delete an existing fleet, simply check the box next to the fleet and then select the

▢ Delete Checked Rows button. And, as on the other tabs, there is a ▢ button

as well as an ▢ button. Selecting ▢ will take you back to the Home screen

STEPS FOR COMPLETING "IDENTIFY FLEETS" SCREEN:

1. On the Home screen, select the 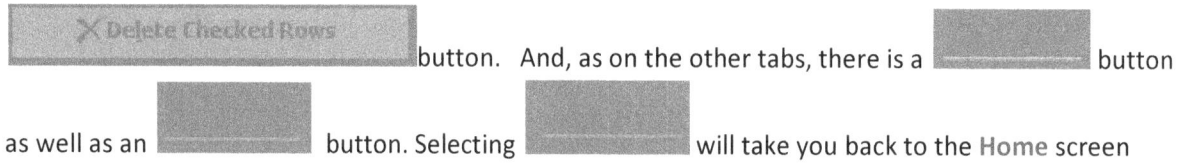 Characterize your Fleets button to display the Fleet Characterization screens.

2. Confirm that the Fleet (Partner Name) that appears automatically is correct and appropriate for your fleets. If not, make changes in the field under the heading "Fleet (Partner Name)."

3. Enter the "Fleet Identifier" for your first (or only) fleet.

4. Enter additional fleets as needed:

    a. To enter another fleet, select the ▢ Add Another Fleet button.

    b. To delete a fleet, select the box next to the row you wish to delete, and then select the ▢ Delete Checked Rows button.

    c. Once you have confirmed or modified the Partner Name and specified the Fleet Identifiers, the full Fleet Names will be displayed on the screen to the right, exactly how they will be displayed on the SmartWay website.

5. To proceed, select the Fleet Details tab at the top, or simply select the ▢ button at the bottom of the screen.

6. Before moving on, a popup screen will appear asking you to verify that you are satisfied with your fleet name(s).

7. ***Verify that the fleet name(s) show exactly what you want customers to find in the SmartWay Carrier Data File.***

    8. Select **OK** to proceed to the next screen. You may return to this screen later to revise your fleet name(s) if necessary.

"FLEET DETAILS" SCREEN OVERVIEW

The Fleet Details Screen is shown below:

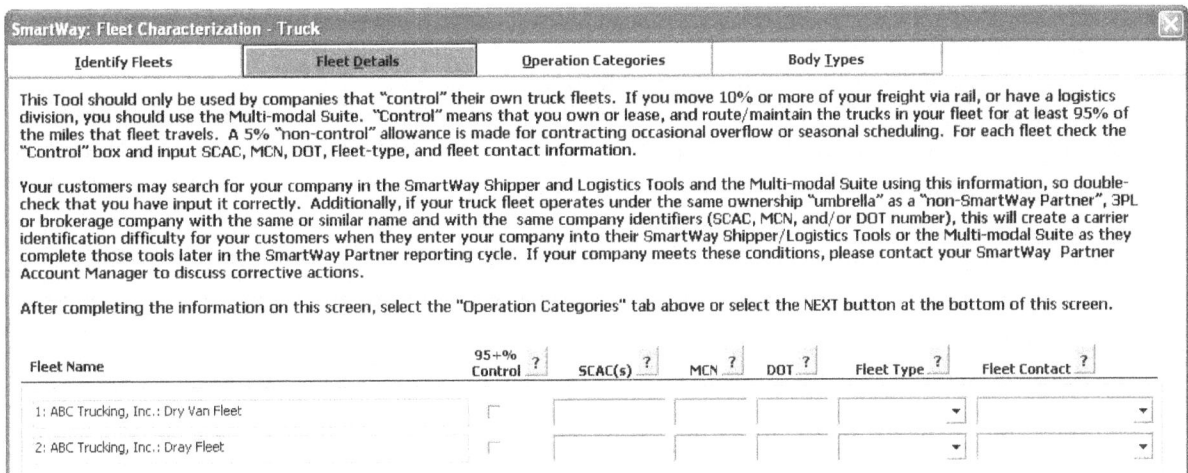

Figure 30: Fleet Details Screen

The Truck Carrier Tool is designed only for fleets over which you have total control. The Fleet Details screen asks you to:

1) Verify and declare that you have control,
2) Add identifying information to help SmartWay Shippers and Logistics Company Partners find your fleets, and
3) Identify the contacts for each of the fleets you defined in the previous screen.

"Control" means that you operate/route the vehicles, regardless of ownership status. Control includes dedicated fleets that you operate for other parties. If you can actively affect the fuel efficiency of the truck and collect the data necessary on that truck to include in this Tool, you have control.

SmartWay understands that there are many organizational/operational variations in the trucking industry and there may be "gray" areas that need further clarification. If you still have questions, you may contact your assigned SmartWay Partner Account Manager or the SmartWay Help Line at 734-214-4767.

NOTE: If you contract out more than 5% of a fleet's operation, the SmartWay Logistics Tool should be used for that fleet.

Once you have confirmed that you control the fleets you have identified, you will be asked to provide further data that helps customers identify your fleets. The Truck Carrier Tool asks you to provide three types of industry identifier codes for each your fleets.

- **Standard Carrier Alpha Codes (SCACs)** are unique 2-4 alphabetic character codes used by the transportation industry to identify transportation companies. SCACs are assigned by the National

Motor Freight Traffic Association, Inc., (NMFTA). If you cannot remember your SCAC(s), contact NMFTA before proceeding. You can find NMFTA contact information at http://www.nmfta.org/Pages/ContactUs.aspx.

- **Motor Carrier Numbers (MCNs):** are 6 or 7 digit numbers provided by the Federal Motor Carrier Safety Administration.

- **Department of Transportation (DOT) Numbers** are carrier identification number issued to all carriers in the U.S. by the Federal Motor Carrier Safety Administration, and can be up to 7 digits in length.

Entering SCACs, MCNs, and DOT number is optional; however, if you have this information you are encouraged to supply it to make sure that SmartWay Shippers and Logistics companies can find you.

You will also need to establish the Fleet Type for each of your fleets, which is defined as the service type for your fleet. There are two options accepted by the Tool—"For-Hire" and "Private/Dedicated." If your company has only one fleet, your "Fleet Type" selection will reflect your company's operations as a whole. If there are multiple fleets, each will have its fleet type defined separately.

Finally, you will identify a Fleet Contact for each fleet. This contact should be one of the contacts already identified in the Contact Information section as the contact for each fleet.

NOTE: A drop-down menu in the Tool will supply this information; if there is a contact for the fleet that is not already listed in the Contacts worksheet, you will need to go back to the Company and Contacts screen to add the required contact information.

STEPS FOR COMPLETING "FLEET DETAILS" SCREEN

1. For each fleet, if you control over 95% of the operation of the vehicles (weighted by miles) check the box labeled "95+% Control."

2. Enter SCAC, MCN, and DOT identifiers (optional) for each fleet listed. If you have a single fleet that has multiple SCACs, enter all of them into the SCAC field, and separate them with commas.

   - While it is not required to enter SCAC, MCN, or DOT information for each fleet, it will help shippers and logistics companies searching by those parameters in the SmartWay database to easily find your fleet for inclusion in their Tool.

3. Select your Fleet Type (either "For-Hire" or "Private/Dedicated") for each fleet listed.

4. Select a Fleet Contact from the drop-down menu for each fleet listed. If the appropriate fleet contact is *not* listed, go back to the Contact Information screen (see Section 2), and enter that name under "Other Contacts." Then return to this screen to add the name from the drop-down menu.

5. Click the [_____] button or select the Operation Categories tab at the top to proceed to the next section.

"OPERATION CATEGORIES" SCREEN OVERVIEW

The **Operation Categories** Screen is shown below:

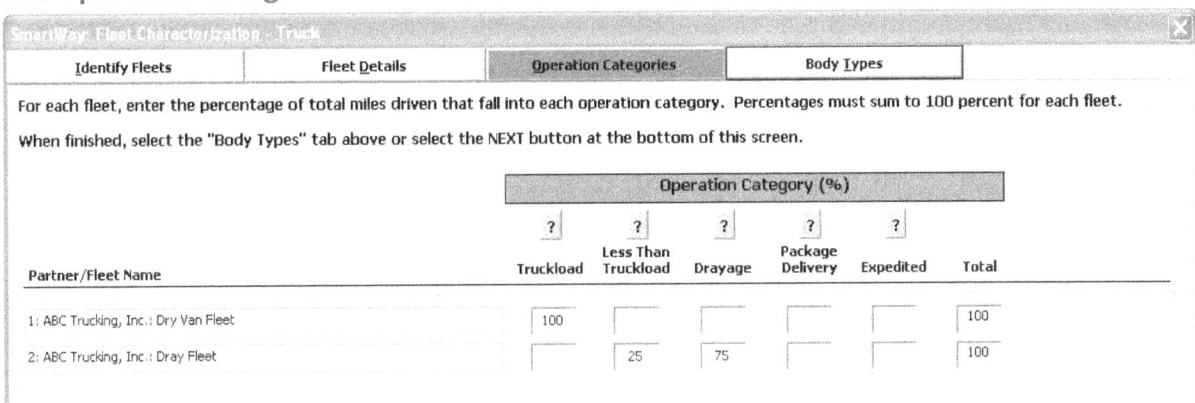

Figure 31: Operation Categories Screen – Example Fleet Inputs

For each fleet, you are asked to estimate the percentage of total mileage that is operated in each of five operation categories defined in the Tool. These Operational Categories are defined below:

- **Truckload (TL)** operations involve movement of large amounts of homogeneous cargo, generally the amount necessary to fill an entire semi-trailer or intermodal container. A truckload carrier is a trucking company that generally contracts an entire trailer-load to a single customer.

- **Less-than-truckload (LTL)** operations involve collecting freight from various shippers and consolidating that freight onto enclosed trailers for linehaul to the delivering terminal or to a hub terminal where the freight will be further sorted and consolidated for additional linehauls.

- **Drayage (Dray)** operations are predominantly associated with port or railhead connections where freight is picked up and moved to a transfer facility or transport mode terminal. These moves are often short in nature, but can be longer depending on specific situations.

- **Package delivery (PD)** operations are characterized by residential or business package pickup/delivery consisting primarily of single or small groups of packages. Package delivery does not include larger scale pickup/delivery operations that are more properly characterized as LTL operations.

- **Expedited operations** are time-sensitive freight shipments, with trucks typically waiting on stand-by.

You will be asked to enter the percent of each operational type based on approximate mileage. This percentage calculation does not need to be exact; however, it should be a reasonable representation of your fleet's operations.

NOTE: You should define your fleet's operations based on your customers' ability to choose them. For example:

- If a fleet is a mix of TL and LTL, you will indicate the percentages of each.

- If customers can choose to hire your TL fleet, your LTL fleet, or your dray fleet separately, then each should be regarded as a separate fleet.

STEPS FOR COMPLETING "OPERATION CATEGORIES" SCREEN:

1. For each fleet identified, estimate the percentage of mileage each fleet spends in the five defined operation categories and enter your estimates in the fields provided. Leave the field blank if no mileage is associated with that operation category for that fleet.

2. Click the [   ] button or select the Body Types tab at the top to proceed to the next section.

"BODY TYPES" SCREEN OVERVIEW

The Body Types screen is shown below:

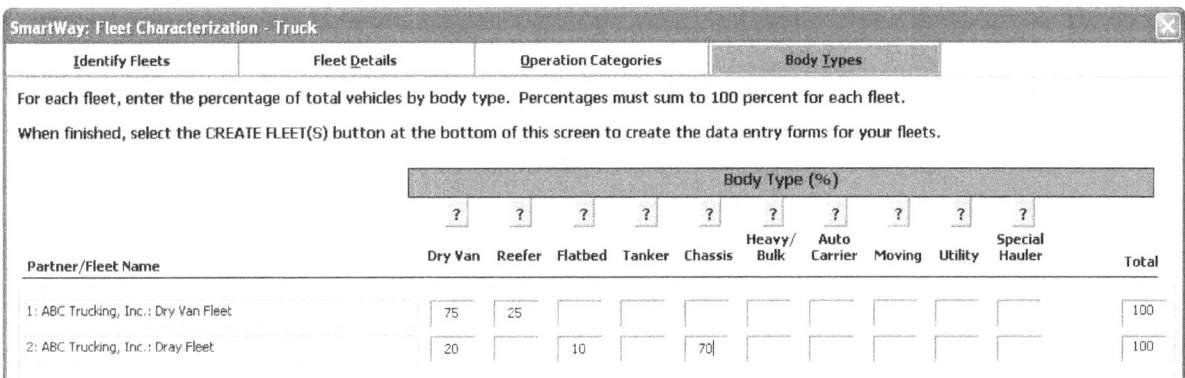

Figure 32: Example for Body Types Screen

Body Type categories as defined in the Truck Carrier Tool are:

- Dry van
- Refrigerated (Reefer)[3]
- Flatbed
- Tanker
- Intermodal chassis containers (pooled and owned)
- Heavy/Bulk hauler
- Auto carriers
- Moving
- Utility[4]
- Special hauler (e.g., Hopper, Livestock, and other specialized carriers)

STEPS FOR COMPLETING "BODY TYPES" SCREEN:

1. For each fleet, enter an estimate of the percentage of fleet mileage associated with each body type. The percentages specified can be approximate, based on vehicle populations. The percentages for each fleet must sum to 100%.

NOTE: If you specify activity for Special Haulers, a **Describe** button will appear next to the cell entry. You must select this button and provide a text description of your specialty haulers.

2. Once you are sure your information is input correctly, you may select the [   ] button at the bottom of the page. You will be prompted then to confirm that you have identified all of

---

[3] If you specify reefer body types in your fleet you must also provide your estimated reefer fuel use in the Activity section of the Tool.

[4] The utility category encompasses class 2b to 8b vehicles that do not carry typical commercial freight. Examples include garbage, recycle, service, work, dump, landscape, cement, bucket, boom, ambulance, armored, fire, farm, wrecker, and other similar trucks. Because these trucks do not carry traditional freight payload, the user should self-define their payloads so as to make the emissions per payload efficiency useful to the user. SmartWay will not use the emissions per payload results for the utility category. Users may experience yellow or red warning labels on the Activity screen due to the unique nature of utility "payload." In the case of red alerts, simply input text defining your special conditions in the required text boxes that appear.

your fleets. Return to the Identify Fleets screen if needed, otherwise select [OK] and you will automatically be returned to the **Home** screen.

Section 4: Select Fleet for Data Entry (providing activity and fuel consumption information for your fleets)

In Section 4 of the Truck Carrier Tool, you will enter detailed activity and fuel consumption data for each of the fleets you identified and characterized in Section 3.

All Truck Carriers should enter the most recent 12 months of data into the Tool. If you are submitting for the first time and do not have a full year of operational data, you must collect <u>a minimum of three months' data</u> for input to submit a tool and join SmartWay. You will be required to submit a full year's data in the follow year's submission.

You will select and enter information about each fleet separately. To begin entering data for a fleet, you must first review the fleet's status in the Section 4 status window.

FLEET STATUS REVIEW OVERVIEW

Once you have returned to the Home screen, you will see all of your fleets listed in the window below item # 4: Select Fleet for Data Entry.

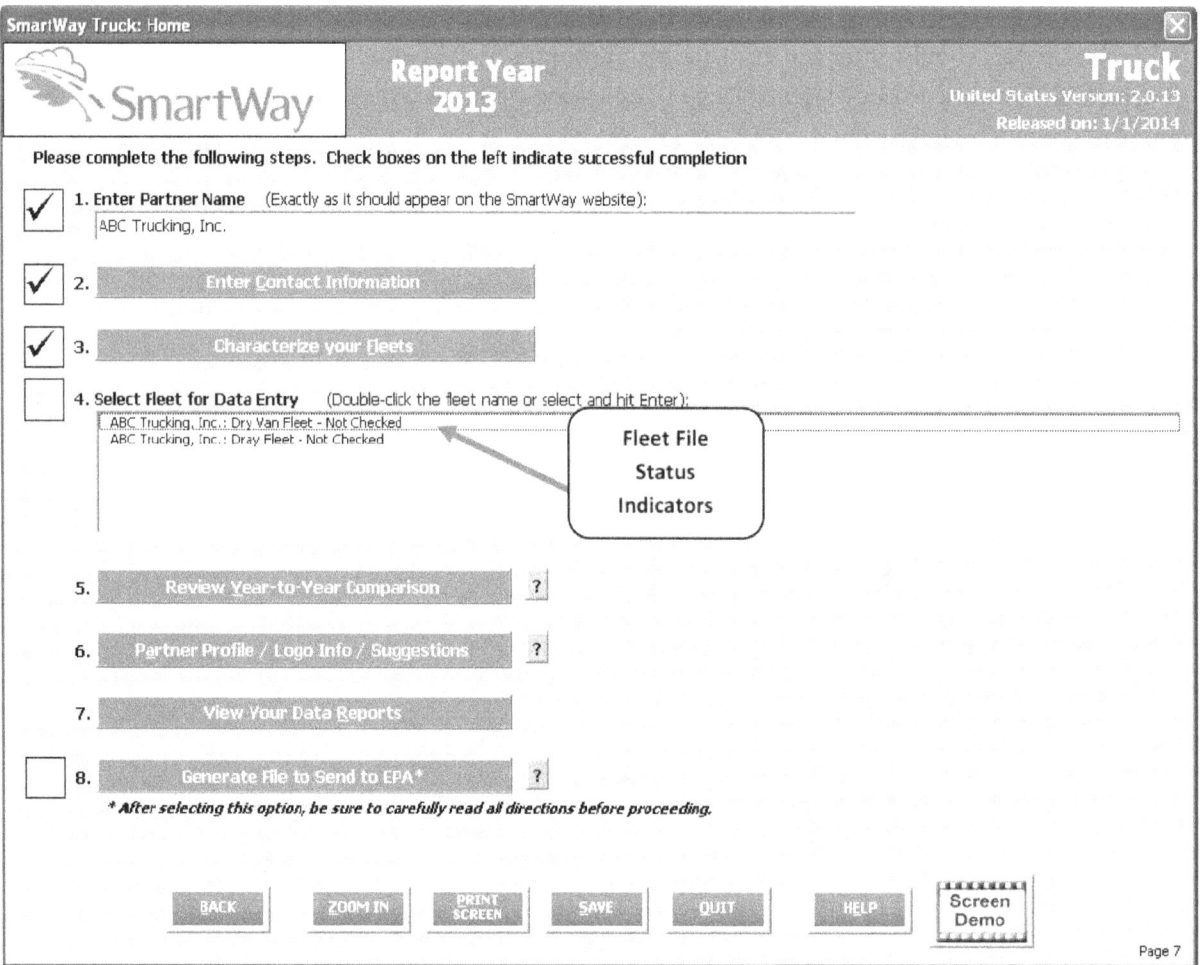

Figure 33: Home Screen – Fleet Status Prior to Activity and Fuel Data Entry

A status message appears after each fleet, indicating whether or not the data entry for that fleet is complete. There are three possible status messages:

- **Not checked** - Data has not been entered yet.
- **Incomplete** - Some data is still missing and/or inconsistent.
- **Complete** - All data requirements have been met and validation has occurred.

In addition to the status messages above, you may also see one of two qualifiers: "Errors" or "Warnings."

- **Errors** prevent you from generating internal metrics reports (under the "View Your Data Reports" section of the Tool), and **must be addressed** before you can submit your Tool to EPA. However, with errors, you will be able to use the **Year-to-Year Comparison Report** to help identify missing data or otherwise clarify uncertainties by referring to previous year submissions.

- **Warnings** will still allow you to run internal metrics reports and submit your data to EPA. **However, it is strongly recommended that you carefully review each warning message before sending your data to EPA** so that you can anticipate questions that may come from a partner account manager (PAM) as a result of your data being outside the expected ranges. The method of addressing errors and warnings is described for the various input screens in the following sections.

To add data to a particular fleet file, highlight the fleet name and then double-click. You will then proceed to the Tool data entry screens.

STEPS FOR SELECTING A FLEET TO REVIEW:

1. Using your mouse, select and highlight the name of the fleet for which you wish to enter data.

2. Double click the name; you will then be taken to the General Information data entry screen for that fleet.

"GENERAL INFORMATION" SCREEN OVERVIEW

The General Information screen is shown below:

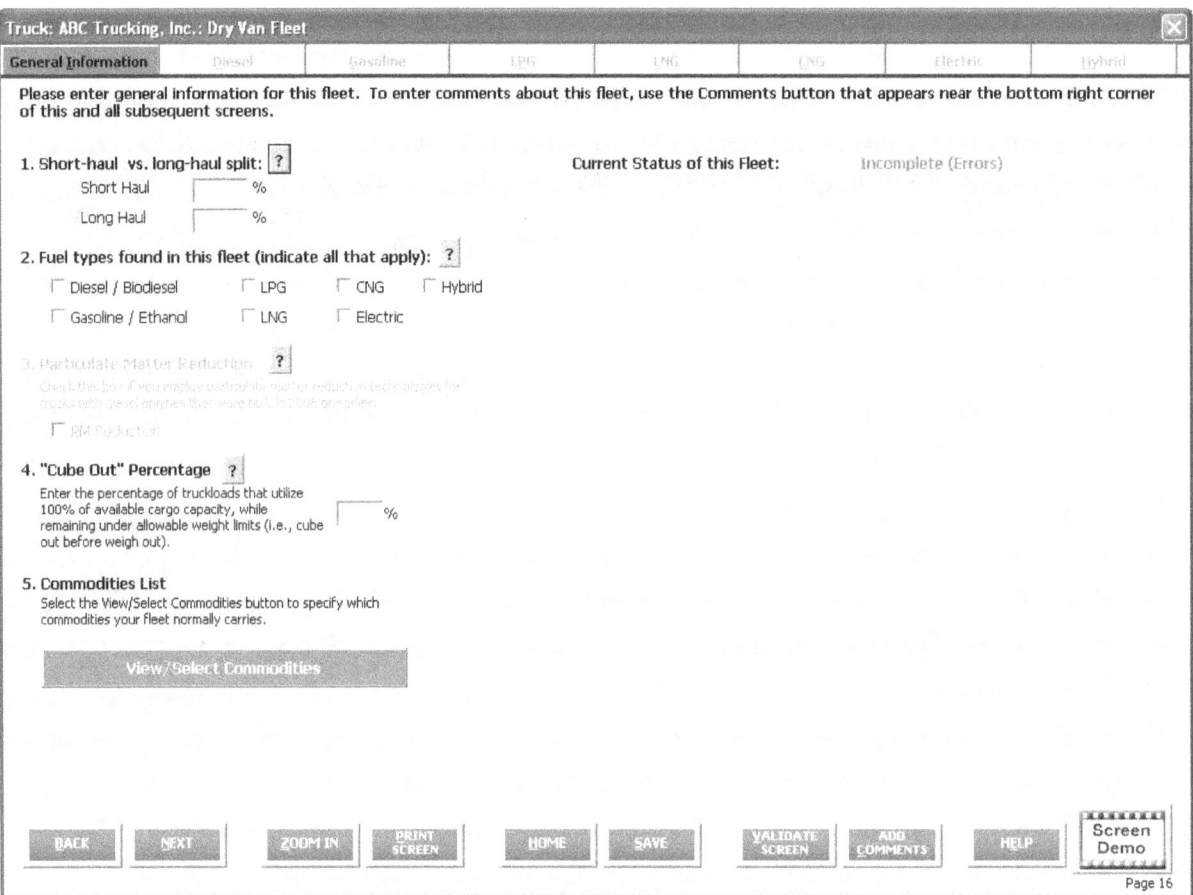

Figure 34: Sample Data Entry Screen for First Example Fleet– General Information

This screen asks you to provide information for six key indicators, including fuel types used.

The first section asks you to determine your short haul split vs. long haul split. This requires you to estimate the percentage of your fleet's operations that are short haul or long haul. A short haul is defined as any haul less than 200 miles; a long haul is defined as any haul in excess of 200 miles. Percentages should be reasonable estimates.

Next, you must select the fuel types found in this fleet. This requires you to define all fuels the fleet used from a list of:

    a. **Diesel/Biodiesel:** petroleum diesel and/or biodiesel made from any renewable feedstock

    b. **Gasoline/Ethanol:** conventional and reformulated gasoline, including blends of 10% ethanol (E10) and 85% ethanol (E85)

    c. **Liquefied Petroleum Gas (LPG):** also known as LP Gas, Liquid Propane Gas, and propane

     d. **Liquefied Natural Gas (LNG)**

     e. **Compressed Natural Gas (CNG)**

     f. **Electric:** vehicles exclusively using batter electric power

     g. **Hybrid:** vehicles with hybrid electric or hydraulic hybrid powertrains, using either gasoline or diesel

NOTE: Once you check these boxes, the appropriate fuel type tab (along the top of the screen next to the General Information tab) will become active.

If you select the Diesel/Biodiesel box, the grayed-out Part 3: Particulate Matter Reduction section will become active.

You will then be asked to determine whether you will need to enter data for Particulate Matter Reduction. Check the Particulate Matter Reduction box only if you have truck engines that are 2006 model year or earlier and are equipped with diesel retrofit particulate matter control devices (i.e., diesel oxidation catalysts (DOCs), particulate filters, or closed crankcase ventilation (CCV)).

Next you will be asked to determine what is known as your "Cube-Out" Percentage. Your "cube out" percentage is the percentage of trailers using 100% of their available cargo capacity while remaining within allowable weight limits. **You must provide a reasonable estimate.**

Next you will go to the [View/Select Commodities] **button to view a list of potential shipment types** and select all of the ones that you move with this fleet.

You also have the option of choosing to participate in SmartWay's Port Dray Program. This voluntary program recognizes Partners for reducing diesel emissions from port drayage trucks. If you have a fleet (or fleets) with 75% or more of operation in the Dray Operation Type category, your fleet is eligible to participate in SmartWay's Port Drayage Program. Appendix A to this document provides the data entry requirements for participation in this program as well as details regarding Dray Program Scoring.

Once you have completed all of the **General Information** screen items, you can select the [   ] button to make sure you have filled out everything on this screen properly. You can also select

[   ] to check your data entries across all screens for the given fleet. Any data entry gaps or inconsistencies will be identified by the Tool.

However, if validating the fleet, note that you will receive additional validation errors unless you have completed the data entry for *all* screens.

STEPS FOR COMPLETING "GENERAL INFORMATION" SCREEN

1. Enter your percentage short-haul vs. long-haul operations.

   - Inputting a value in one cell automatically populates the other cell to add up to 100.

2. Check the boxes for the fuel types you use.

   - Once you check these boxes, the appropriate fuel type tab (along the top of the screen next to the General Information tab) will become active. If you select the **Diesel/Biodiesel** box, the grayed-out **Part 3: Particulate Matter Reduction** section will become active.

3. Determine whether you will need to enter data for Particulate Matter Reduction; if yes, check the box.

   - Check the Particulate Matter Reduction box only if you have truck engines that are model year 2006 or earlier and are equipped with diesel retrofit particulate matter control devices (i.e., diesel oxidation catalysts (DOCs), particulate filters, or closed crankcase ventilation (CCV)).

4. Enter the percentage of trailers using 100% of available cargo capacity while remaining within allowable weight limits.

5. Click the ⬛ View/Select Commodities button and select which commodity categories you typically carry. Select all categories that apply to your fleets.

6. ***FOR FLEETS WITH 75% OR MORE OF TRUCKS IN THE DRAYAGE OPERATIONS CATEGORY:*** Check the box if you wish to participate in the Port Dray Program.

7. Select the ⬛ button to make sure you have filled out everything on this screen properly.

   You can also select ⬛ to check your data entries across all screens for the given fleet.

8. Click ⬛ or select the next available fuel tab at the top to begin entering fuel and activity data for this fleet.

FUEL SECTION: "ENGINE MODEL YEAR & CLASS" OVERVIEW

This section looks at truck classes and engine model years. Under each fuel-type tab (Diesel, Gasoline, LPG, LNG, CNG, Electric, and Hybrid) there are as many as four screens requiring data inputs:

1. The Engine Model Year & Class[5] screen,
2. The Activity Information screen,
3. The PM Reduction screen will appear for diesel vehicles _if_ you checked the **Particulate Matter Reduction** box on the General Information screen, and
4. The Port Dray Program screen will also appear for diesel vehicles _if_ you checked the **Port Dray Program** box on the General Information screen.

The Diesel Fuel Types Screen is shown below—this screen defaults to the first available sub-tab, the Engine Model Year & Class screen:

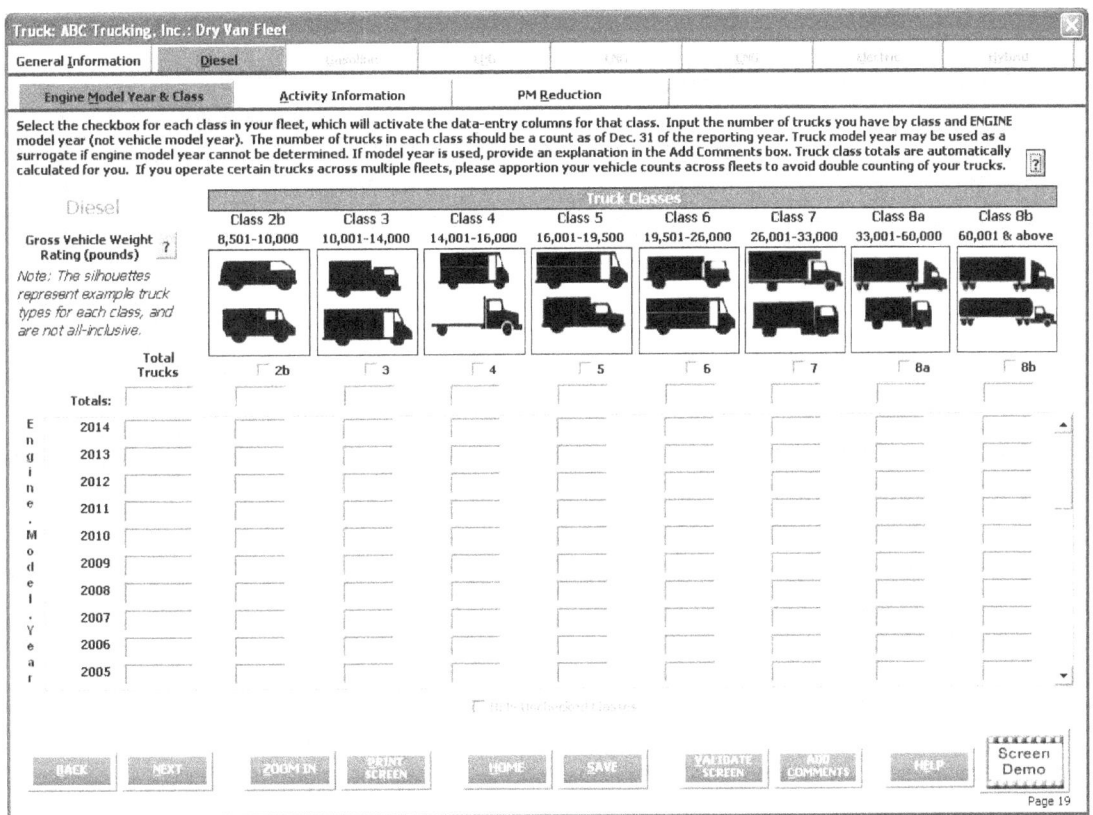

Figure 35: Engine Model Year & Class Screen

NOTE: For each fuel type that you identified on the General Information screen, **you will be <u>required</u> to complete the** Engine Model Year & Class **screen AND the** Activity Information **screen.** If you do not complete these screens for any of the fuels, your fleet will be marked as **Not Complete** on the Home screen and you will not be able to submit your Tool.

---

[5] For Electric, this screen is known as the "Motor Model Year and Class" Screen.

The first step on the Engine Model Year and Class **tab is to identify the vehicles classes you have represented in this fleet.** Begin by selecting the boxes at the top for each of the truck classes you operate in this fleet (i.e., 2b, 3, 4, 5, 6, 7, 8a, 8b). Example truck types are shown in silhouette above the boxes, and additional examples will be listed on the screen when placing the mouse over these images.

When a Truck Class box has been checked, the data column will activate.

Next you must enter the number of vehicles you have in each class, specifying the corresponding **engine model years** (rather than the vehicle (tractor) model years). Use the scroll bar to the right if you need to enter information for model years earlier than 2005.

Totals for each truck class are calculated automatically for you and displayed along the top row. Totals by model year are shown in the left-hand column.

NOTE: If you defined multiple fleets on the Fleet Characterization screen, and if you operate certain trucks across multiple fleets, apportion your vehicle counts across the fleets to avoid double counting of your trucks. For example, if you operate the same 100 trucks across two fleets, with 20% of the truck mileage for the first fleet, then enter 20 trucks for fleet #1, and 80 trucks for fleet #2.

STEPS FOR COMPLETING THE "ENGINE MODEL YEAR & CLASS" SCREEN

NOTE: The following guidance will use the Diesel Fuel sections as an example. Similar procedures are followed for all other fuel types.

1. Select the boxes at the top (i.e., 2b, 3, 4, 5, 6, 7, 8a, 8b) for each of the truck classes you operate in this fleet.

2. Input the number of vehicles you have in each class, specifying the corresponding **engine model years** (rather than the tractor model years). Use the scroll bar to the right if you need to enter information for older model years.

3. Check the box at the bottom of the screen to hide any unused truck classes if you wish.

4. Select [   ] or select the Activity Information tab at top of the screen to proceed to the next section. Before leaving the Engine Model Year & Class screen, you will be prompted to confirm the accuracy of your model year and truck class selections (**Figure 36**). You may review previous years' selections by selecting the **Year-to-Year Comparison Report** on the Home screen.

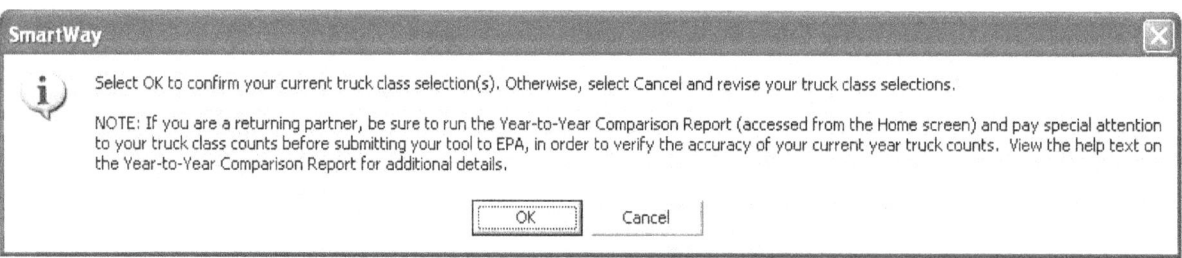

Figure 36: Confirmation of Engine Model Year and Truck Class Selections

NOTE: Class 8a trucks are often mistaken for Class 8bs. If you select to enter data for Class 8a trucks you will be asked to confirm your selection. Please refer to the Gross Vehicle Weight Rating Definitions at the top of the screen before proceeding.

FUEL SECTION: "ACTIVITY INFORMATION" SCREEN OVERVIEW

The Activity Information screen is shown below:

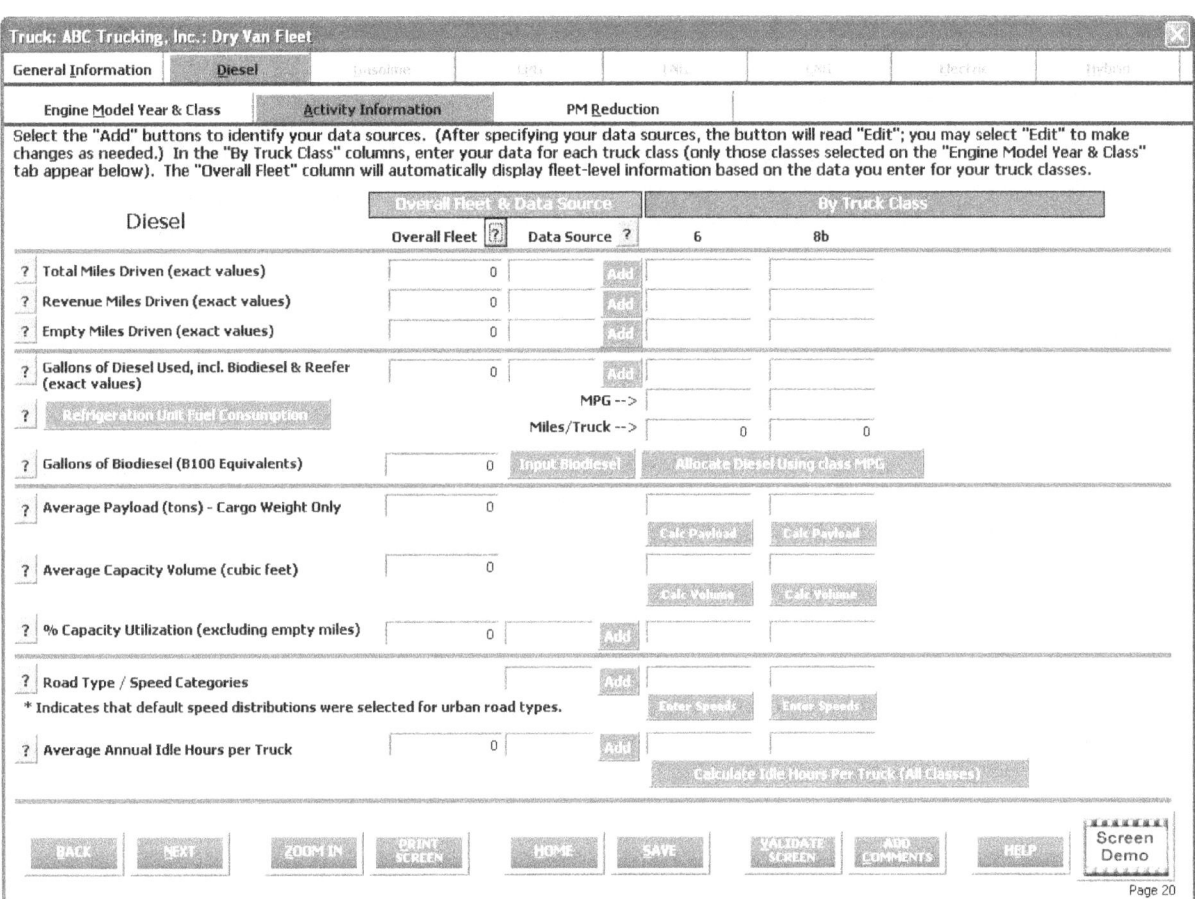

Figure 37: Diesel Vehicles Activity Information Screen

The text at the top left-hand corner of this screen indicates the fuel type for which you are entering data. You will need to enter data for each fuel type you use within your fleet under separate screens, including:

- diesel and biodiesel, entered under the same tab
- gasoline and ethanol, entered under the same tab
- LPG, LNG, CNG, hybrids, and electric trucks each entered under separate tabs

If you selected more than one fuel type for this fleet on the General Information screen, then you will also need to enter data on the other fuel type tabs. For instance, if you operate gasoline vehicles, select the Gasoline Vehicles tab on the main tab bar to enter your data.

*****Alert! Be careful to input data under the appropriate fuel tab!*****

On each fuel type's Activity Information screen, you will see two main sections to input data:

- **Overall Fleet & Data Source** section (with a green header), and
- **By Truck Class** section (with a blue header).

The Overall Fleet data indicates the sum or average (as appropriate) for the fleet across all truck classes you have selected on the Engine Model Year and Class screen. This value is automatically tallied when you enter data in the "By Truck Class" fields (the blue header section.)

MAIN FUEL ACTIVITY DATA POINTS OVERVIEW

The first five activity data points are:

- **Total Miles Driven (exact values):** all miles driven collectively by all fleet vehicles in a specific class

- **Revenue Miles Driven (exact values):** **all** miles that were charged to a customer account.
  - **NOTE:** If you have a private fleet that does not track revenue miles for internal cost accounting, set revenue miles equal to total miles - DO NOT SET REVENUE MILES EQUAL TO ZERO.
  - This information is not used to calculate your SmartWay carbon score, but may be used determine an adjustment factor for shippers' carbon inventories.[6]

- **Empty Miles Driven (exact values):** the total number of empty miles traveled by your fleet.
  - Empty is defined as zero cargo.
  - **NOTE:** If you enter zero (0) empty miles, you **must** provide text explaining why you have no empty backhauls by double-clicking on the cell.
  - The number of empty miles will not affect your SmartWay score. Companies in categories with high empty miles, such as tanker operations, **will not be negatively affected** by high empty-mile values.

- **Gallons of Diesel Used, including biodiesel and reefer (exact values):** all gallons of fuel used by your fleet in the **past 12-month reporting period.**
  - This includes any gallons with biofuels (biodiesel for diesel vehicles, ethanol for gasoline vehicles).
  - It also includes gallons used for refrigeration, bunk heaters, yard moves, or any other gallons directly attributable to transportation. Refrigeration (reefer) gallons must also be reported separately.
  - It does NOT include gallons used in heating buildings, forklifts, or other non-transportation sources.
  - **NOTE: If total gallons of diesel fuel used are known for the fleet but not by class, the Fuel Allocator Worksheet will help you apportion your total fuel use into class categories.**

- **OPTIONAL: Gallons of Biodiesel or Gallons of Ethanol (for diesel and gasoline fuel tabs, respectively):** the B100 equivalents for biodiesel or the E100 equivalents for ethanol used by the fleet

It should be noted here that all values for mileage, gallons, and other inputs should reflect EXACT values rather than rounded estimates. Failure to provide exact values will trigger warnings upon validation, reminding you to input exact values.

---

[6] Since shippers determine their carbon inventory based on revenue miles charged by their carriers, shippers collectively must also include the non-revenue miles in their carbon inventory. Thus shippers will be charged the carbon for the non-revenue miles that carriers have to travel. This value will be calculated on an industry basis, and will not affect individual carriers.

For each of these four data activity points, you must specify a data source.

To add data source information, select the [Add] button under the **Overall Fleet & Data Source** section to specify where you obtained your data for each row.

Use the pull-down menus on the Data Source Description popup screens to specify the source of information for each Data Type (listed in red at the top of the form). Most of the data source information you specify is assumed to apply across all vehicle classes in your fleet.

A helpful reference table with available sources is included in Appendix B.

In addition to the general data source, further detail must be provided regarding *how* the particular data was collected (e.g., via GPS or odometer readings).

After selecting from the "Data Source" and "Source Detail" drop-down menus, you may be required to provide additional information regarding the way you collected your data and associated calculation methods, as appropriate. You should provide enough detail to confirm data validity and reasonableness.

Once the data is entered, the [Add] buttons will read **Edit**, allowing you to change your choices later.)

NOTE: "Data Source" descriptions for payload and volume information are specific to each truck class, and are entered through the Payload and Volume Calculators (select [Calc Payload] and [Calc Volume] to access; see **Figure 38**).

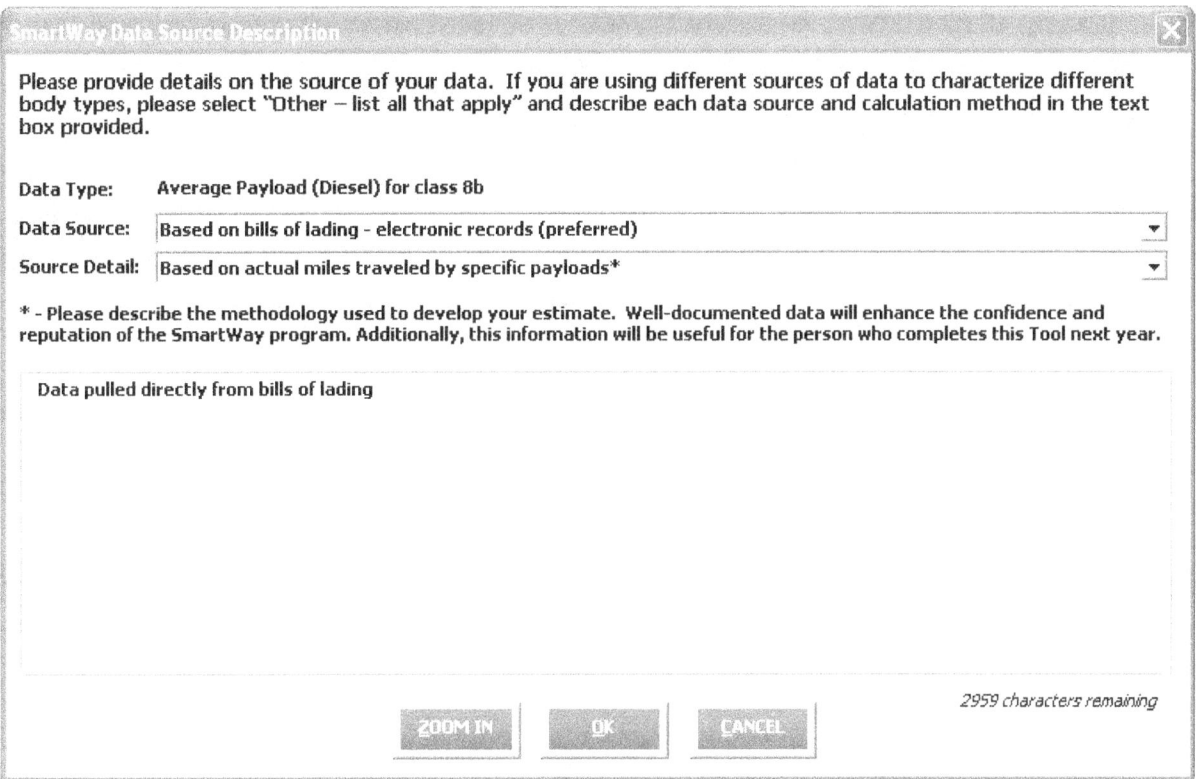

Figure 38: Example Data Source Description Screen

Once your data sources have been selected, you can then enter the requested activity information for the reporting period for each truck class in the white empty cells to the right (under the blue heading "By Truck Class."

STEPS FOR COMPLETING THE MAIN FUEL DATA POINTS ON THE "ACTIVITY INFORMATION" SCREEN

1) Select the **Add** button under the **Overall Fleet & Data Source** section to specify where you obtained your data for each row. For each data type, specify the general type of data source.

2) Enter the exact value for total miles driven collectively by this fleet by vehicle class. Include all out-of-route, positioning, empty, and other miles driven.

3) Enter the exact value for the revenue miles—the number of miles your fleet drove that were charged to a customer account.

4) Enter the exact value for the total number of empty miles traveled by your fleet.

5) Enter the exact value for all the gallons of fuel used by your fleet in the **past 12-month reporting period**, including any gallons with biofuels (biodiesel for diesel vehicles, ethanol for gasoline vehicles).

6) If you specified any use of refrigerated body types under the **Fleet Characterization** section, the ▨ button will appear underneath the Gallons of Fuel Used section. Select this button and provide your best estimate of your refrigerated fuel use in gallons for each

truck class. Diesel, LPG, and electric trucks are assumed to use diesel reefers, gasoline trucks to use gasoline reefers, and CNG/LNG trucks to use CNG reefers. Enter zero for any truck classes/fuel types that do not utilize reefer units. If you do not know your reefer fuel use at the truck class level you may enter a total value which will be distributed across your truck classes proportional to your vehicle fuel consumption entries.

Figure 39 below presents the Reefer Fuel Data Entry screen. Enter your reefer fuel consumption in gallons for each vehicle class if possible. If one or more vehicle classes do not include reefers, enter 0 gallons. The Truck Tool assumes diesel fuel is used for reefers associated with diesel, LPG, and electric trucks, while gasoline reefers are associated with gasoline trucks, and CNG reefers are associated with CNG and LNG trucks.

If you do not know your reefer fuel consumption at the vehicle class level check the box and enter your total reefer fuel consumption in the Total cell. This total value will then be allocated across all vehicle classes proportional to the total vehicle class fuel consumption amounts.

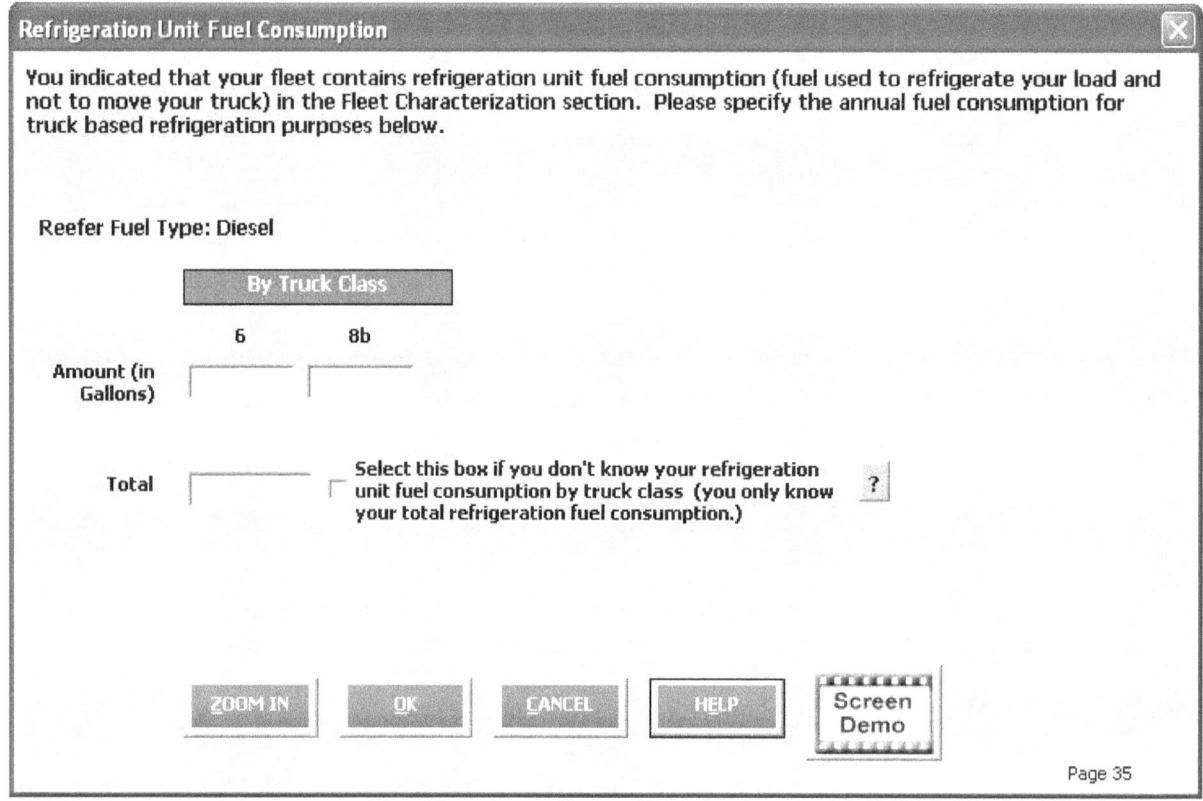

Figure 39: Example Refrigeration Unit Fuel Consumption Data Entry Screen

7) NOTE: On the Electric Vehicle Activity screen, "fuel inputs" are expressed in kWhrs rather than gallons.

"FUEL ALLOCATOR WORKSHEET" OVERVIEW

The Truck Carrier Tool asks you to enter Gallons of Diesel Used by truck class. This information may be entered directly if you have it. However, if you do not, but you do know total fuel use and MPG by truck class, the Fuel Allocator will assist you in apportioning your fuel use across truck classes.

To use the Fuel Allocator Worksheet, you will select the **Allocate Diesel Using class MPG** button on the Activity Information screen beneath the data entry cells for the "Gallons of Diesel Used, incl. Biodiesel & Reefer (exact values)" section.

Figure 40 shows the Fuel Allocator Worksheet. Note that you must enter total miles for each truck class on the Activity Information screen before you can open this worksheet.

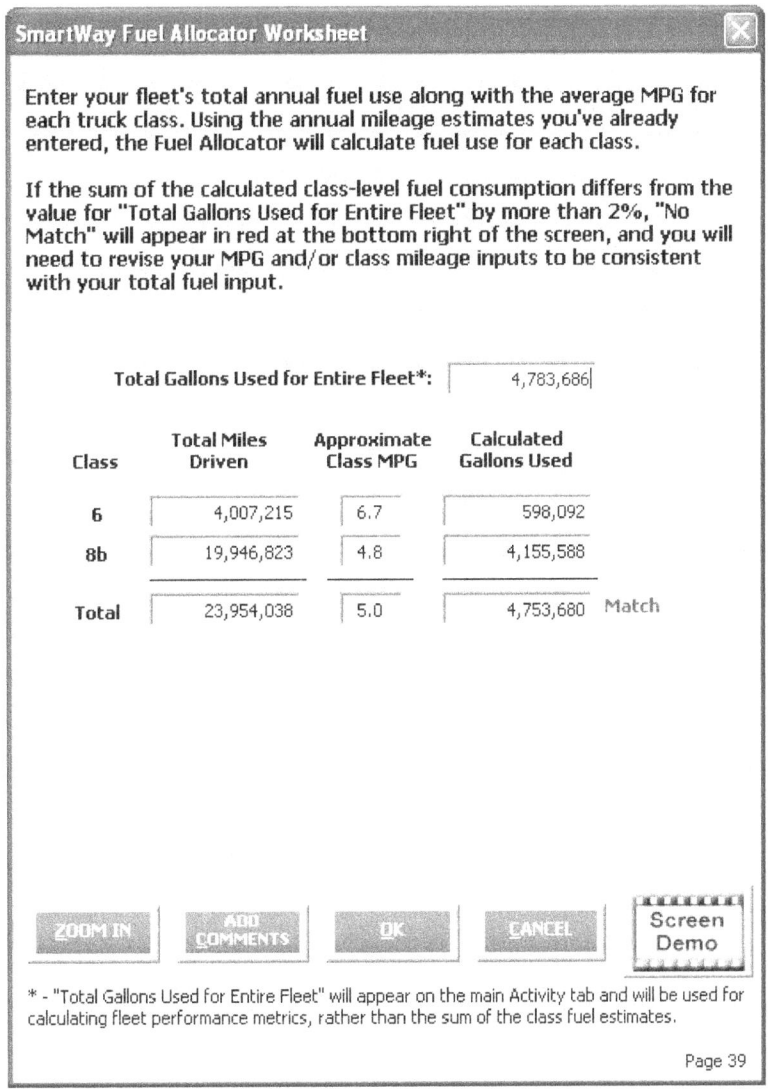

Figure 40: Fuel Allocator Worksheet

The **Fuel Allocator Worksheet** requires you to estimate:

- total fuel consumption for your entire fleet across all vehicle classes, and
- MPG for each truck class (the worksheet for electric vehicles uses miles/kWhr in lieu of MPG).

Only truck classes selected on the **Engine Model Year & Class** screen will be displayed on the **Fuel Allocator Worksheet**. Class-specific fuel consumption levels are estimated using the annual mileage estimates entered on the **Activity Information** screen.

If the sum of the class level estimates is within 2% of the total fuel consumption level entered at the top of the form, then a "**Match**" is indicated on the right side of the worksheet, and you may go back to the **Activity Information** screen by pressing **OK**.[7]

Before returning to the **Activity Information** screen, you will see a popup:

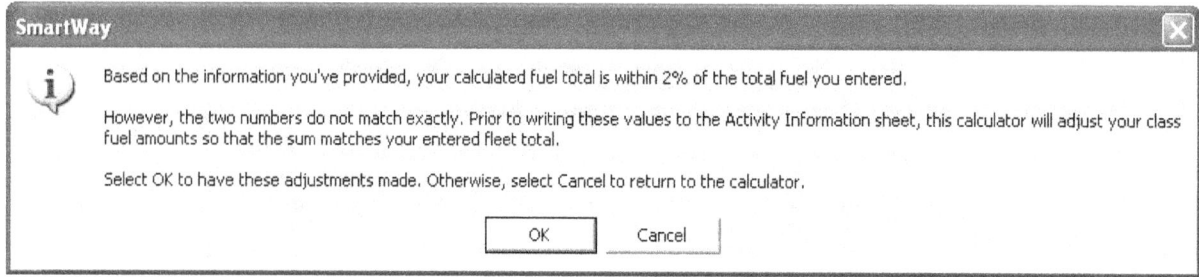

Figure 41: Popup Indicating Fuel Total Match

This popup indicates that the total gallons of diesel you entered and the MPG by truck class that you entered produced a calculated total fuel use that is slightly different. If you select **OK**, this calculator will adjust the by class fuel amounts to a sum matches your entered fleet total.

If "No Match" is indicated, you must adjust your total fuel consumption estimate and/or your class level MPG estimates until a "**Match**" is obtained.

Once you complete the Fuel Allocator Worksheet, total gallons will be summed and displayed in the **Overall Fleet** column on the **Activity Information** screen.

There are two other sections within the "Gallons of Diesel Used, incl. Biodiesel & Reefer (exact values)" section, MPG and Miles/Truck, described below.

MPG: If you used the Fuel Allocator Worksheet to determine your fuel usage, the MPG fields will show the MPG estimates that you entered into that form. If, instead, you entered gallons of fuel used by class, the Tool will generate MPG estimates for you. This field is provided as a useful validation check for users. If the Tool finds an out-of-range MPG value during its routine validations, the MPG field may be

---

[7] Although the calculated and total fuel consumption values are within 2%, the values may not match exactly. Prior to writing the fuel consumption values to the Activity Information screen, the calculator will proportionally adjust your class level fuel consumption estimates to match your entered fleet total value.

highlighted, in which case you may be instructed to double-click on this cell to provide an explanation for high/low values.

 Miles/Truck: Once Total Miles are entered for a given truck class, this value will be combined with the truck count values entered on the Engine Model Year and Class screen to provide a Miles/Truck value, which is displayed below the MPG value on the Activity Information screen. This field is provided as a useful validation check for users, and is checked by the Tool's validation routines for reasonableness. If the Tool finds an out-of-range miles-per-truck value during its routine validations, the field may be highlighted, in which case you may be instructed to double-click on this cell to provide an explanation for high/low values.

These two fields do not involve direct data entry; they are populated as a result of Tool calculations and are based on your entries for total miles by truck class and total gallons by truck class (either directly entered or allocated using the Fuel Allocator Worksheet).

STEPS FOR COMPLETING THE "FUEL ALLOCATOR WORKSHEET"

1. Select the [Allocate Diesel Using class MPG] button on the Activity Information screen.

2. Enter total gallons of diesel for all truck classes.

3. Enter MPG estimates for each truck class.

4. If "Match" appears in blue next to the "Calculated Gallons Used" field, click [   ].

5. If "No Match" appears in red next to the "Calculated Gallons Used" field, adjust the MPG values as necessary to get a match, and then click [   ].

(OPTIONAL) "BIODIESEL BLEND WORKSHEET" OVERVIEW

If your fleet consumed any amount of biodiesel during the reporting period, you will have access to a Biodiesel Blend Worksheet, shown in **Figure 42**. (A similar worksheet is available for ethanol use under the Gasoline tab – see below.)

The Biodiesel Blend Worksheet asks you to specify the biodiesel volume used by your fleet in gallons by blend level (e.g., B20 = 20% biodiesel / 80% conventional diesel), for the reporting period. The calculator will automatically display total gallons, the weighted average blend value (across all blends), and the B100-equivalent volume at the bottom of the screen, based on your entries.

The Truck Carrier Tool will assume that these gallons are allocated proportionally across all vehicle classes (weighted by gallons of diesel used) in order to calculate emissions.

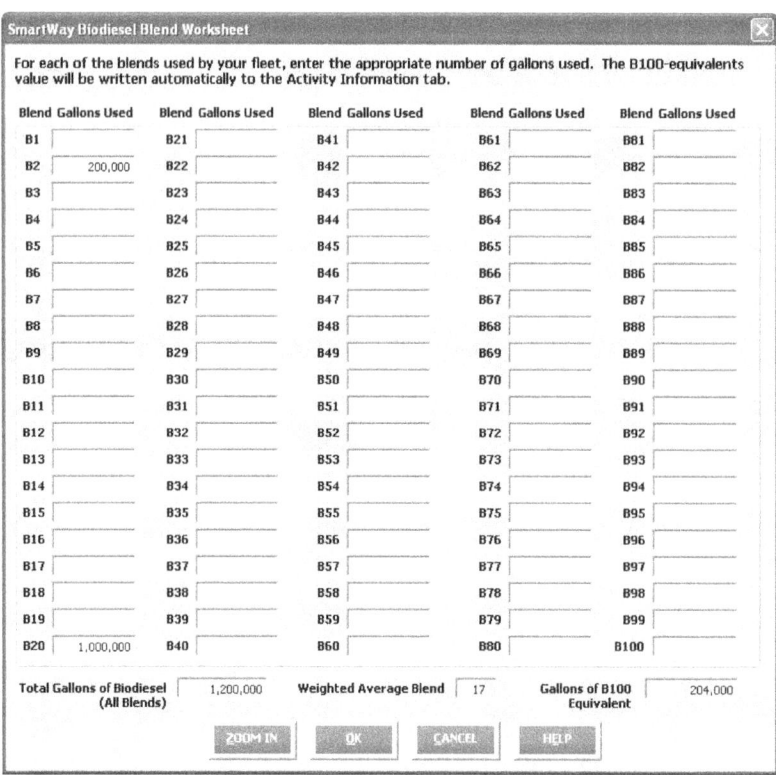

Figure 42: Biodiesel Blend Worksheet

STEPS FOR COMPLETING THE "BIODIESEL BLEND WORKSHEET "

1. If this fleet has used biodiesel, select the ⬚ Input Biodiesel ⬚ button and specify your biodiesel volumes by blend level in the Biodiesel Blend Worksheet.

2. For each of the blends used by your fleet, enter the appropriate number of gallons used.

3. Select **OK.**

(OPTIONAL) "ETHANOL BLEND WORKSHEET" OVERVIEW

If your fleet consumed any amount of ethanol during the reporting period, you can use the Ethanol Blend Worksheet, shown in **Figure 43,** to determine your ethanol usage figures.

If you know the volume of ethanol used by your fleet, specify the volume in gallons for each blend level (E10 and/or E85), ensuring that the total volume specified does not exceed the gasoline gallon data entry provided on the Activity Information screen. (Enter zero for both values if your fleet does not use one of these blends.)

Alternatively, if you do not know the volume of ethanol used by your fleet, select "**Use national average for ethanol usage (9.05%)**." In either case, the total volume of ethanol at each blend level will be allocated across the different truck classes in your fleet in proportion to the total gallons of gasoline used.

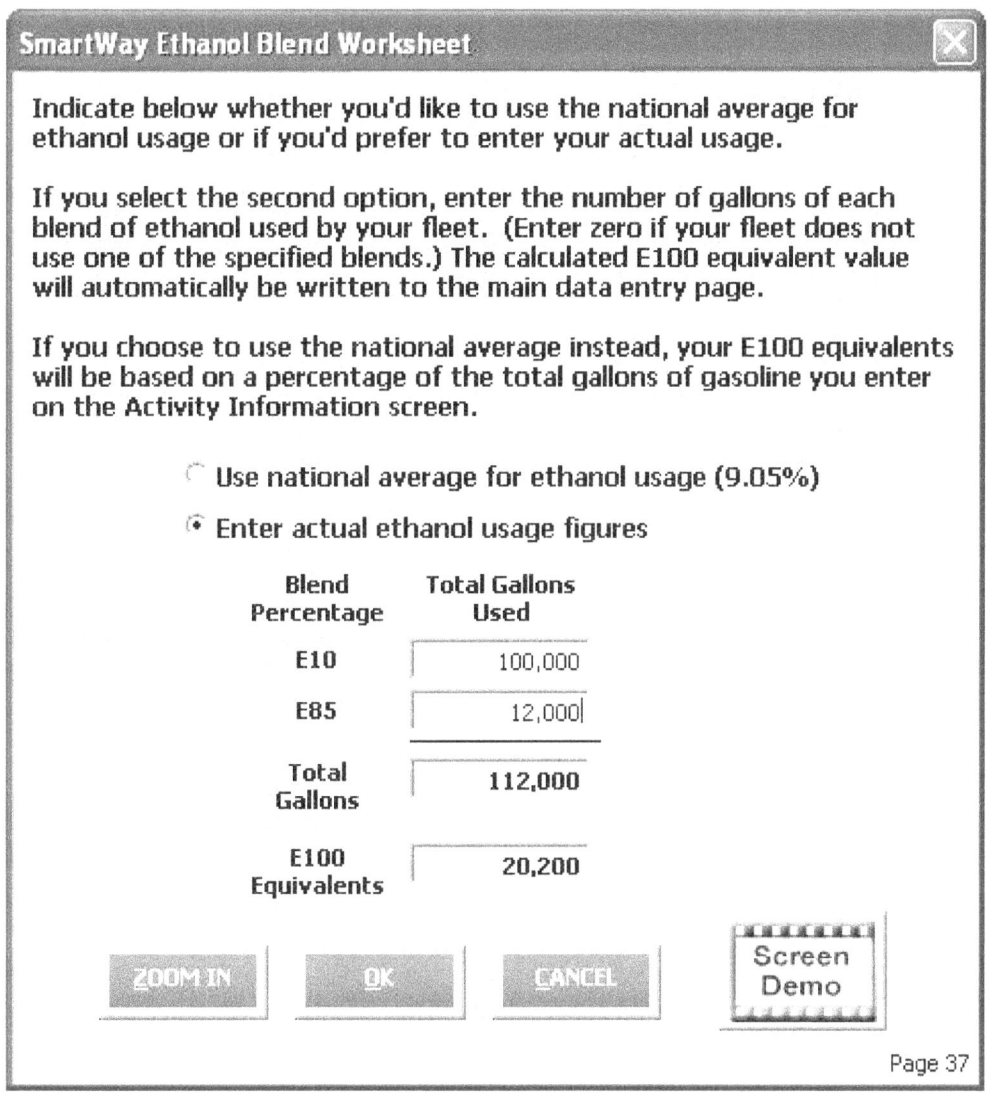

Figure 43: Ethanol Blend Worksheet

1. If this fleet has used ethanol, select the [Input Ethanol] button and specify your ethanol volumes by blend level (E10 or E85) in the Ethanol Blend Worksheet.

2. For each of the blends used by your fleet, enter the appropriate number of gallons used.

3. Select **OK.**

PAYLOAD AND CAPACITY OVERVIEW ON THE "ACTIVITY INFORMATION" SCREEN

This section of the Activity Information screen asks to calculate your average payload by truck class using an internal payload calculator.

Average payload calculations are made on a "by truck class" basis. Within a given truck class, you may have many different body types.

For each body type within a truck class, you will need to choose:
- which of the **four available allocation methods** you will use to distribute your trucks' activity to calculate a weighted average payload value by truck class:
  - # miles by class
  - # trips by class
  - % of operation by class
  - # vehicles by class
- whether you are interested in looking at your payload in **units** of short tons (=2,000 pounds) or in pounds
- whether to either **directly calculate an average payload** for this body type (using bill of lading data collected by your company) or to select one of five **average payload ranges** provided in the Tool (which reflect typical industry payload ranges).

The values entered for each body type provide the basis for calculating a weighted average payload for the class as a whole. For example, by specifying the mileage associated with each body type, the respective miles per year will be used to weight your payload estimates to calculate an average by truck class.

Ideally, you should calculate your average payload data by using payload data from your company's bills of lading records. Note that, when calculating average payload, your payload data should represent the average cargo weight per LOADED trip; empty backhauls should be excluded from the calculation.

For truck classes where you have precise payload data, select either electronic or manual bills of lading (with data based on actual miles traveled or trip-weighted), or another data source (for which you must provide a text description).

For truck classes for which you DO NOT have precise payload data, select the "Used ranges provided by the calculator" as your data source.

The Payload Calculator displays up to three screens.

The first screen is labeled "Step 1: Getting Started." See **Figure 44** below.

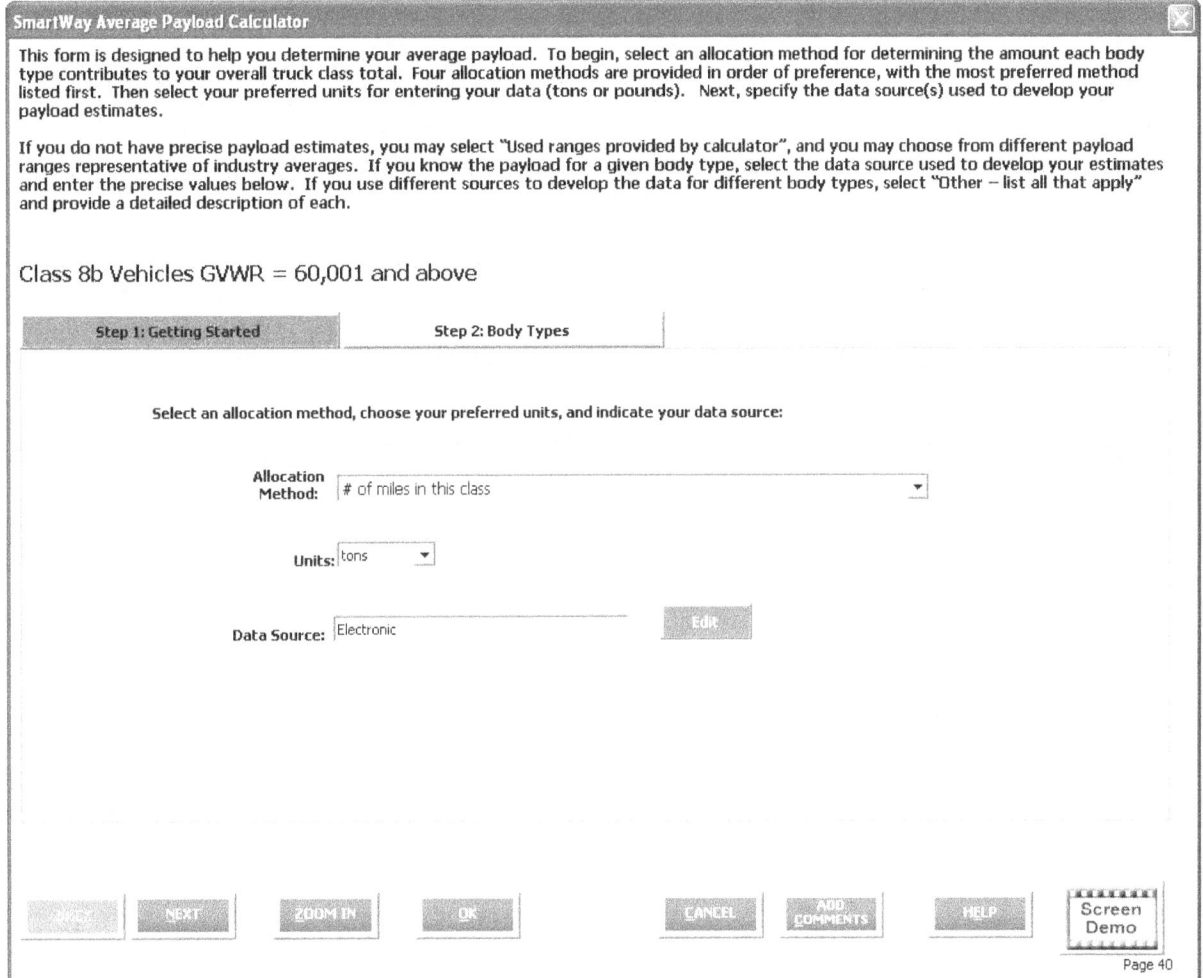

Figure 44: Step 1 of Payload Calculator

On this screen, you will be asked to make three selections:

- Allocation Method
- Units
- Data Source

The first choice you will make is the payload allocation method. The method selected will be used to allocate your payloads across multiple truck body types within a given truck class. This allocation allows the Tool to calculate a weighted average payload value by truck class.

There are four payload allocation methods available:
- # miles by class
- # trips by class
- % of operation by class

- # vehicles by class

Next you must choose between either short tons (2,000 pounds) or pounds as your preferred weight metric or "Unit."

Next you will choose your Data Source(s) for your payload entries.

At this point you will proceed to the next tab, labeled "Step 2: Body Types" to enter payload information specific to each body type for the given truck class.

Figure 45 presents the "**Step 2: Body Types**" tab of the Average Payload Calculator.

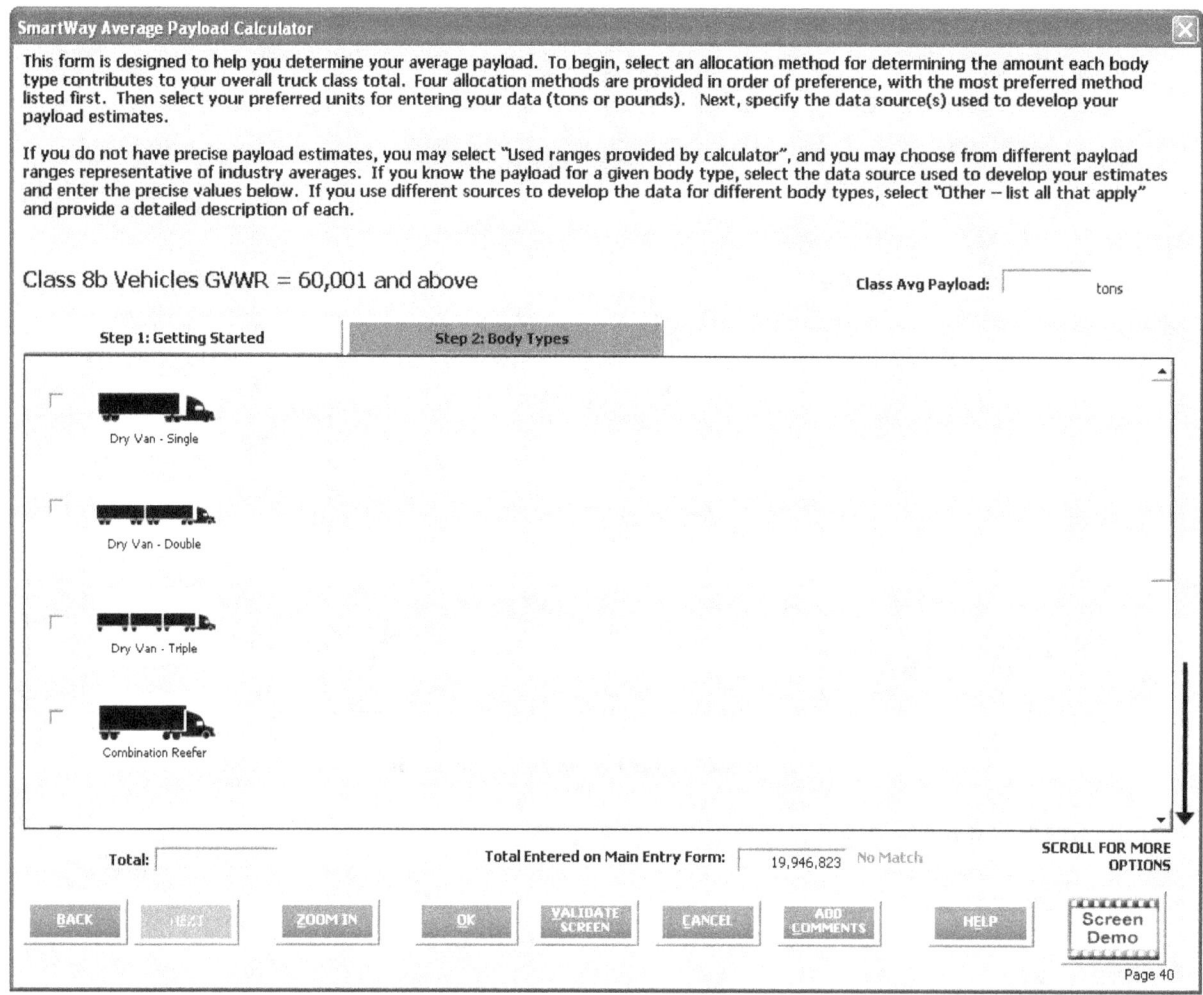

Figure 45: Step 2 of Payload Calculator

Use this screen to select the applicable body types for the truck class, specify the payload allocation information for each body type (e.g., number of miles traveled), and the average payload information.

If you specified the "Used ranges provided by the calculator" option under "Data Source" in Step 1, select from pre-defined Ranges 1 – 5 (1 being the lowest payload range values available and 5 being the highest).[8] Use the scroll bar to the right to display other body types.

If you need to use pre-defined ranges for some body types and precise values for other body types, select the "Other – list all that apply" Data Source option in Step 1.

If Range 1 (extreme low) or Range 5 (extreme high) values are selected, the value will be highlighted in red, and explanations must be provided in the text box to the right summarizing the reasons for the unusual payload value (e.g., you may explain that a flatbed fleet is regularly used to transport heavy construction equipment).

If you select Range 2 (moderately low) or Range 4 (moderately high), the selection will be highlighted in yellow and a warning will appear. Text explanations are optional for Ranges 2 and 4.

If an exact value entry falls outside the expected (Range 3) values, an explanation will be requested.

Figure 46 shows Step 2 of the Average Payload Calculator worksheet for Class 8b vehicles. (Similar worksheets are provided for each truck class.)

---

[8] Specific range values were determined based on reported industry payload distributions, and are discussed in detail in the Truck Tool Technical Documentation.

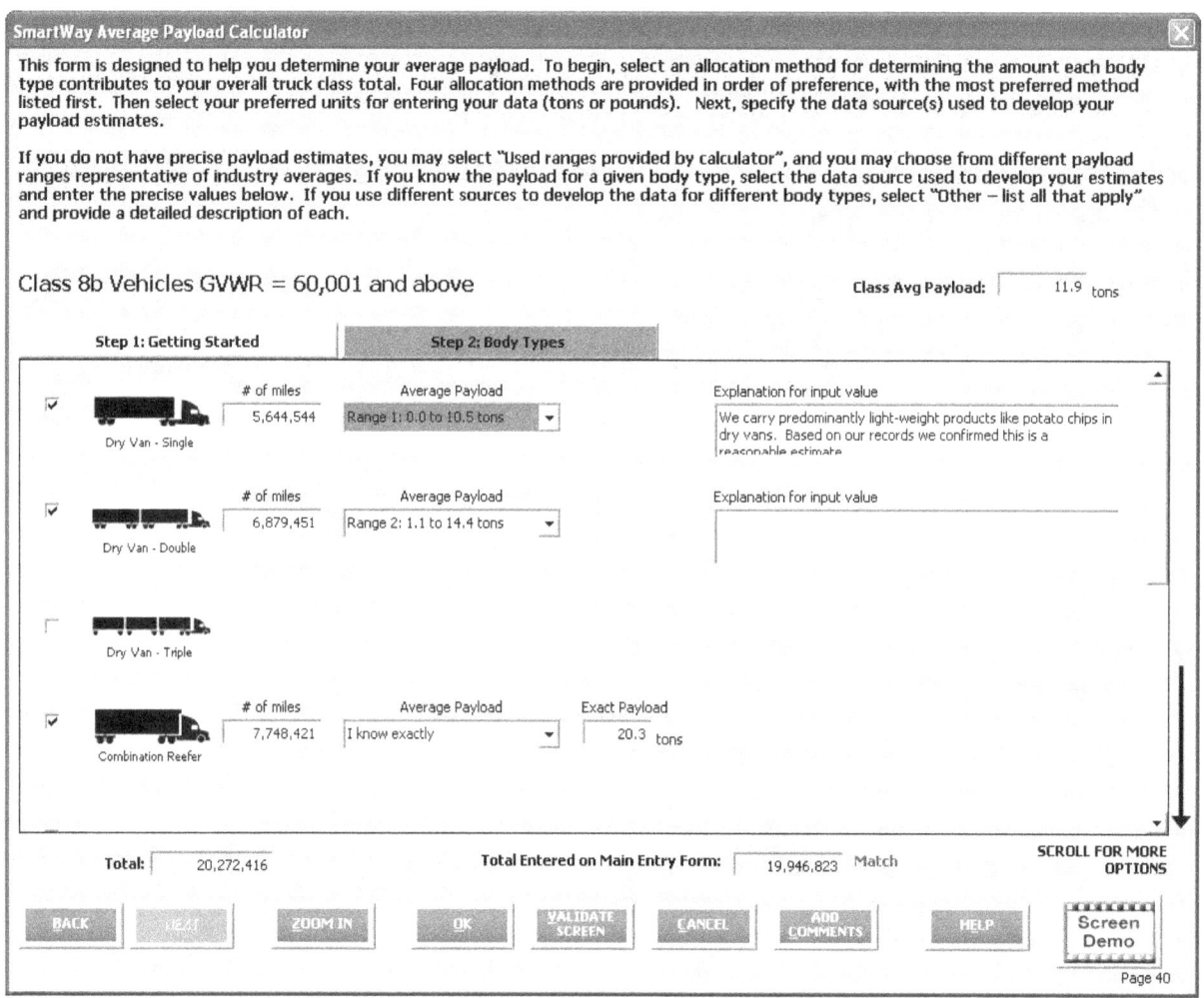

Figure: 46: Step 2 of Payload Calculator Worksheet – Class 8b Vehicles Data with Explanations

Once the mileage and payload ranges or values are entered for each body type, the worksheet validates the entries. If the sum of the by-body type values is within 2% of the class level mileage totals from the Activity Information screen, a "Match" is indicated and the user may proceed with the rest of the data entry. Otherwise "No Match" is indicated and the mileage values must be revised until "Match" is indicated. A validation is also performed if "# of "was selected in Step 1 as the allocation method, although in this case the match must be exact. Independent matching is not performed for "# of trips" or "% of operation allocation methods."

If any of the values entered in the Average Payload Calculator received Red or Yellow warnings, the weighted class average payload value shown in the Activity Information screen will be highlighted in yellow as a reminder.

Once specified for each *class*, the *fleet*-average payload will be calculated (weighted by the number of miles per class) and displayed in the **Overall Fleet** column on the Activity Information screen.

EXAMPLE:

ABC Trucking, Inc. has a fleet with Class 8b trucks. This fleet consists of three body types: dry van single trailer, dry van double trailer, and reefer.

In "Step 1: Getting Started," ABC Trucking, Inc. chooses "# of miles by class" as the allocation method and "short tons" as the unit.

Under "Data Source," ABC Trucking, Inc. selected "Other." This allows the company to use ranges for some data and exact average payload calculations for others.

ABC Trucking proceeds to "Step 2: Body Types," and selects the checkboxes next to the three body types present in the Class 8b truck fleet (dry van single trailer, dry van double trailer, and reefer).

ABC Trucking, Inc. enters the total number of miles driven with loads (excluding empty miles) for each body type.

Next, the company selects the Average Payload from the drop-down lists.

In the case of the dry vans with single trailers, Range 1 is selected. Because this is considered an unusual result, the Tool highlights this payload, in red and ABC Trucking, Inc. MUST enter an explanation to proceed.

In the case of the dry vans with double trailers, Range 2 is selected. Because this is considered somewhat unusual, the Tool highlights this payload in yellow, and ABC Trucking, Inc. has the OPTION of entering an explanation to proceed.

In the case of the combination reefers, ABC Trucking, Inc. knows EXACTLY what the average payload value is and enters it in the field provided.

STEPS FOR COMPLETING THE PAYLOAD SECTION ON THE "ACTIVITY INFORMATION" SCREEN

1. Enter your average payloads into the Tool for each truck class by selecting the [Calc Payload] button.

2. Under the "Step 1: Getting Started Tab," select your activity Allocation Method.

3. Select your Units, from the drop-down.

4. Click the [Add] button next to "Data Source." A Data Source Description box will appear.

5. Using the drop-down menus, select your data source and details and enter any comments about your data source.

6. Click [   ].

7. Click the "Step 2: Body Types" tab or [   ].

8. Check the box(es) next to the body types used in this Truck Class.

9. For each body type selected, enter the activity associated with this body type.

10. For each body type selected, use the drop-down to select a range OR enter the exact payload if available. Make sure to include any pallet and packaging weight in your payload estimates.

11. Validate the screen to determine if there are any errors to correct; if yes, correct the errors and/or enter comments using the [   ] button to explain your data inputs.

12. Click the [   ] button to perform a final validation check and return to the Activity Information screen.

AVERAGE CAPACITY VOLUME (CUBIC FEET) WORKSHEET OVERVIEW ON THE "ACTIVITY INFORMATION" SCREEN

Volume refers to the total cargo-carrying capacity of your vehicles, not the utilized space on the vehicles.

Volumes are specific to each major body type/configuration, and can be expressed in cubic feet or twenty-foot equivalent units (TEU).

The volume calculator worksheet for truck classes 2b through 7 follow the same format as the average payload calculators, with various body type selections available.

Similar to payload data entry, you may enter your calculated volume capacity information by selecting the "Determined using company records" for the "Data Source" drop-down option within the Average Capacity Volume Worksheet.

Alternatively, you may select default values for each body type by selecting the "Used defaults provided by calculator" for the "Data Source" drop-down option.

Finally, if you need to use different sources of data to characterize different body types, select the "Other – list all that apply" for the "Data Source" drop-down option within the calculator.

The volume calculator worksheet for classes 8a and 8b reflect a range of different, standard trailer and container configurations. **Figure 47** shows an example worksheet for Class 8b vehicles.

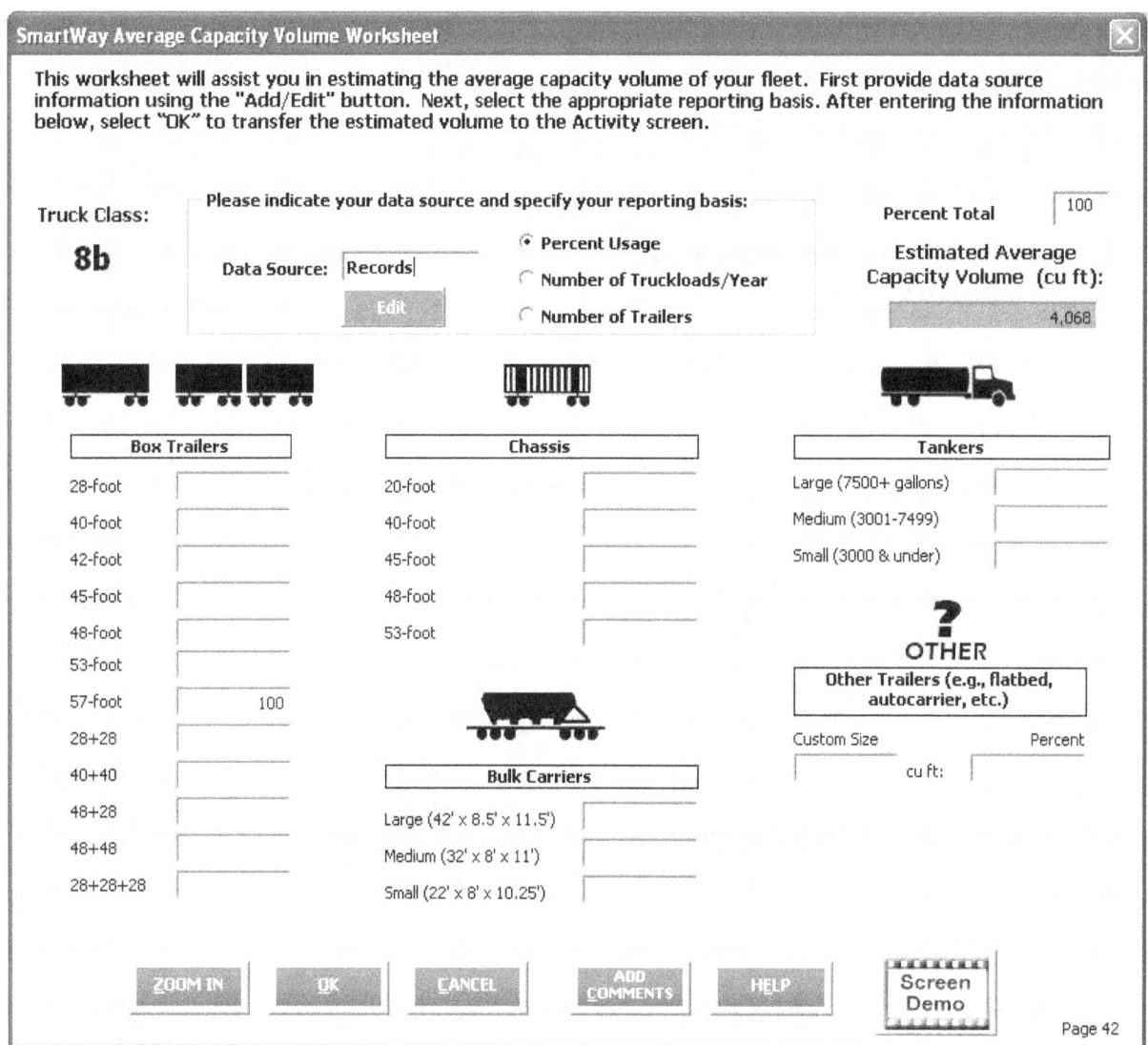

Figure 47: Screenshot of Average Capacity Volume Worksheet

First you must select the "Data Source" for your volume capacity data and then specify your reporting basis from the three available choices (Percent Use, Number of Truckloads per year, Number of Trailers). Next you will specify the relative distribution of your trailer/container sizes for your truck class as a whole.

 If you select "Percent Usage" as your reporting basis, the values entered on this screen must sum to 100. A running total is provided on the upper right of the screen.

If you use "Other" trailer types not specified, include both the average volume of the trailers in cubic feet under "Custom Size," as well as the reporting basis value. Once complete, the estimated average volume weighted across all trailer/container types will be displayed in the upper right of the screen in cubic feet. This value will be written to the Activity Information screen for the specific truck class. **NOTE: If you operate flatbeds enter your trailer information here under the "Other" category.**

Make sure trailer/container type selections are consistent with the body types specified in the Payload Calculator and Fleet Characterization screens; inconsistencies will be flagged during validation.

Once specified for each class, the fleet-average capacity volume will be calculated (weighted by the number of miles per class) and displayed in the Overall Fleet column on the Activity Information screen.

STEPS FOR COMPLETING AVERAGE CAPACITY VOLUME (CUBIC FEET) WORKSHEET ON THE "ACTIVITY INFORMATION" SCREEN

1. Click the [Calc Volume] button to open the Capacity Volume Calculator screen.

2. Click [Edit] to enter your data source for average capacity volume data.

3. Select one of the three reporting basis options using the radio check boxes.

4. Enter the requested metrics for the trailers, containers, tankers, bulk carrier, or other trailers used in this truck class.

5. Select [   ] to return to the Activity Information Screen.

PERCENT (%) CAPACITY UTILIZATION SECTION OVERVIEW ON THE "ACTIVITY INFORMATION" SCREEN

Percent capacity utilization refers to the percentage of trailer capacity that is used for all hauls in a given fleet.

Percent capacity utilization applies *only* to loaded (non-empty) miles. For most carriers, this will reflect your loaded volumetric fill rate.

However, there are a few exceptions:
- LTL carriers should estimate the weighted-average fill rate over the span of operations.
- Flatbed haulers should estimate fill rate based on deck area covered.
- Auto carriers should estimate based on percent of car slots filled.
- Drayage operations may select the data source "Unknown to dray carriers - use industry average" option, in which case the industry average capacity utilization value will be auto-populated for you.

***Do not factor in empty miles for percent capacity utilization. ***

Once specified for each class, the fleet-average percent capacity utilization will be calculated (weighted by the number of miles per class) and displayed in the **Overall Fleet** column.

STEPS FOR COMPLETING THE % CAPACITY UTILIZATION SECTION ON THE "ACTIVITY INFORMATION" SCREEN

1. Next to the "% Capacity Utilization (excluding empty miles)" data entry box, click the Add button. A Data Source Description box will appear.

2. Using the drop-down menus, select your data source and details and enter any comments about your data source.

3. Click _____ to return to the Activity Information screen.

4. Enter your "% capacity utilization" for each truck class.

ROAD TYPE/SPEED CATEGORY SECTION OVERVIEW ON THE "ACTIVITY INFORMATION" SCREEN

"Road Type/Speed Category" is a ratio of total mileage by class allocated to highway/rural driving, and to urban driving less than 25 mph, 25-50 mph, and over 50 mph.

Data for completing this screen should be available from your vehicles' electronic control modules (ECM).

NOTE: If you do not know your speed distribution for urban areas, you may check the box labeled "Populate the urban driving fields with default values," and default percentages will be calculated based on data from EPA's MOVES model,[9] adjusted for the Highway/Rural Driving percentage specified.

(For further details, see the Truck Carrier Tool Technical Documentation available at http://epa.gov/smartway/partnership/trucks.htm)

STEPS FOR COMPLETING THE ROAD TYPE / SPEED CATEGORY SECTION ON THE "ACTIVITY INFORMATION" SCREEN

1. Click [Add] next to the data entry box next to "Road Type/Speed Categories." A Data Source Description box will appear.

2. Using the drop-down menus, select your data source and details and enter any comments about your data source.

3. Click [   ] to return to the Activity Information screen.

4. For each truck class, select [Enter Speeds] to open the ""Road Type/Speed Categories" box.

5. Enter the requested metrics for Highway or Rural Driving and Urban Driving for each of the speeds listed (if applicable). If you only know Highway or Rural driving, you can select the box next to "Populate the urban driving fields with default values" to complete the screen.

6. Click [   ] to return to the Activity Information screen.

---

[9] NOx and PM emission rates for diesel, gasoline and E10 vehicles are obtained from EPA's MOVES 2010b emissions model, expressed in grams/mile. Emission rates for LPG, CNG, and LNG are adjusted from gasoline values. Hybrid vehicle emissions are set equal to their corresponding conventional gasoline and diesel counterparts, less the idle emissions component. Finally, electric vehicle CO_2, NOx and PM emission rates are based on national average electricity generation unit estimates (in grams per kWhr). See the Truck Tool Technical Documentation for additional details on emission rate estimation.

AVERAGE ANNUAL IDLE HOURS PER TRUCK SECTION OVERVIEW ON THE "ACTIVITY INFORMATION" SCREEN

The **Average Annual Idle Hours per Truck**[10] is calculated using the Idle Hours Calculator. To calculate idling hours, you will need to enter the number of hours your trucks idle per truck per day on average, along with the average number of days each truck is in service, on average, per year.

Figure 48 shows the Idle Hours Calculator screen.

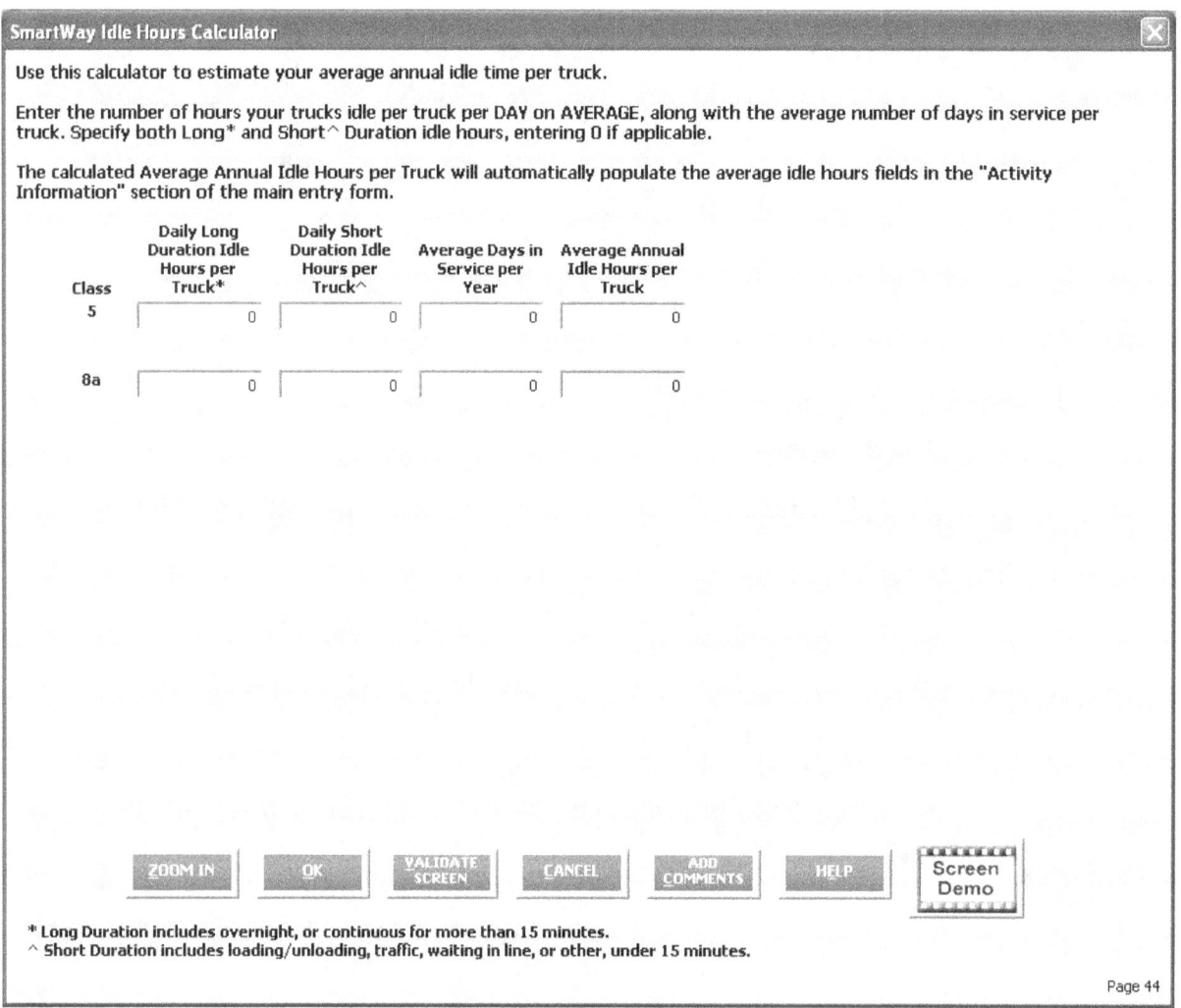

Figure 48: Screenshot of Blank Idle Hours Calculator

Separate values should be input for short-duration (less than 15 consecutive minutes) and long-duration (greater than 15 minutes) idling.[11] After you have entered long and short duration idling hours and

---

[10] Idle hours are not applicable to hybrid or electric trucks, so the Idle Hour Calculator is not provided on those screens.

[11] NOx and PM emission rates are different for short and long-duration idling, reflecting the varying engine loads and operating temperatures for these events. The SmartWay emission calculations distinguish between these events for Class 8b trucks using MOVES model outputs, although different emission factors may be integrated in future versions of the model for all truck classes.

average days in service for each truck class, and the calculator will present idle hours for each class will be presented on the Activity Information screen.

In addition, fleet-average idle hours across all truck classes are then calculated (weighted by vehicle counts) and displayed in the **Overall Fleet** column.

STEPS FOR COMPLETING THE AVERAGE ANNUAL IDLE HOURS PER TRUCK SECTION ON THE "ACTIVITY INFORMATION" SCREEN

1. Click [Add] next to the data entry box next to "Average Annual Idle Hours per Truck." A Data Source Description box will appear.

2. Using the drop-down menus, select your data source and details and enter any comments about your data source.

3. Click [_____] to return to the Activity Information Screen.

4. Click the **Calculate Idle Hours Per Truck (All Classes)** button to open the Idle Hours Calculator.

5. Enter daily long-duration idling hours per truck, daily short-duration idle hours per truck, and average days in service per year for each vehicle class represented in this fleet.

6. Click [_____] to return to the Activity Information Screen.

7. You have now completed the Activity Information Screen for this fuel type.

8. Click the [_____] button to verify that you have completed the screen properly.

9. After completing the Activity Information screen, if you are using PM reduction equipment select the **PM Reduction** tab at the top of the screen to proceed to the next section.

"PM REDUCTION" SCREEN OVERVIEW

The PM Reduction screen is for fleets that have installed retrofit equipment on pre-2007 engines. You are provided with radio button to select the type of device (DOC, CCV, or PM trap) you have used for your fleet. You are then asked to enter the number of trucks equipped with the device by <u>engine</u> (not vehicle) model year.

To enter data for multiple devices, select the radio button for the next type of device after completing data entry for the initial device, and fill out the new blank fields that appear.

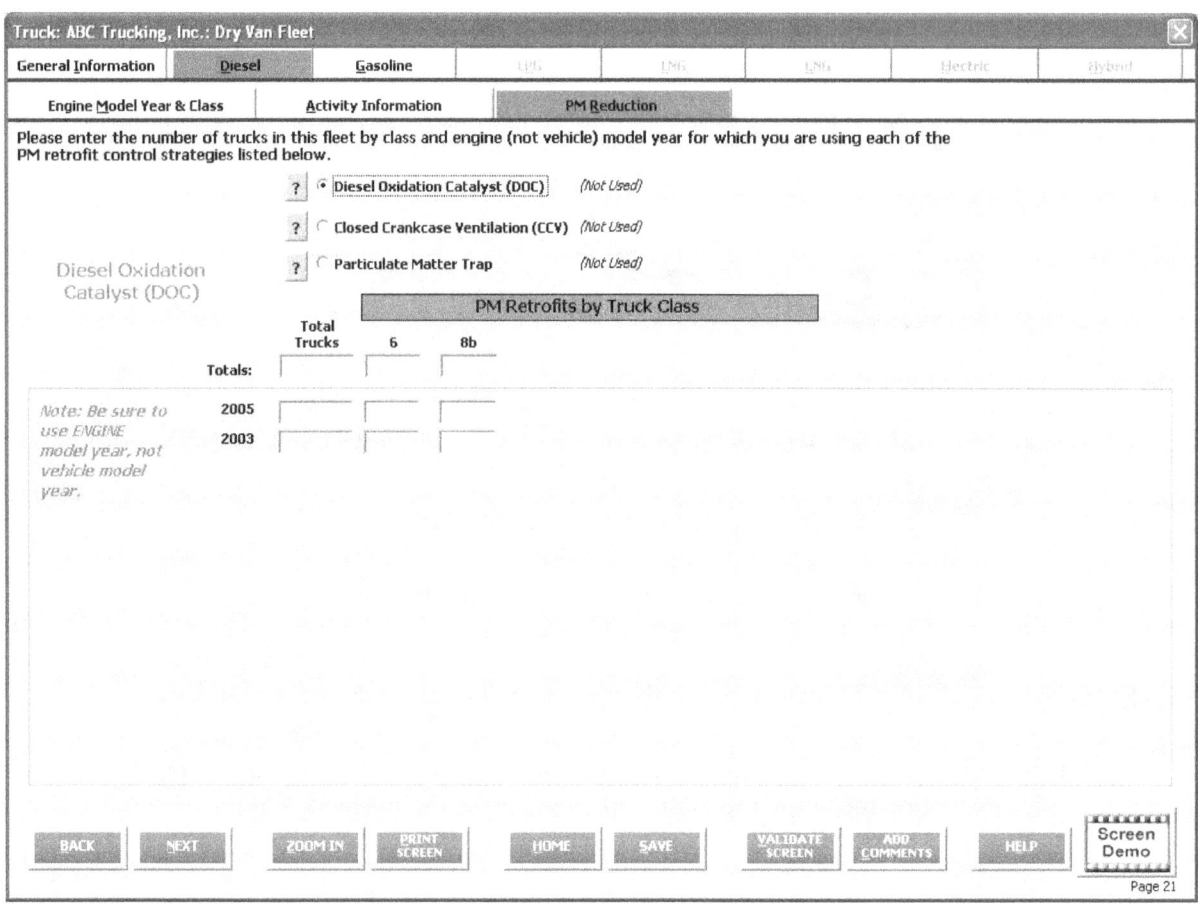

Figure 49: Diesel Vehicles PM Reduction Screen

Totals for any given model year cannot exceed the totals specified on the Engine Model Year & Class **screen.** While CCVs may be installed in combination with either DOCs or PM traps, it is assumed that DOC and PM trap applications are mutually exclusive. As such, **the sum of DOC and PM trap trucks cannot exceed the totals specified on the** Engine Model Year & Class **screen.**

STEPS FOR COMPLETING THE "PM REDUCTION" SCREEN

1. Select the radio button next to a device (DOC, CCV, or PM trap) you have used for your fleet.

2. Enter the number of trucks equipped with the device by <u>engine</u> (not vehicle) model year.

3. If other devices have been used with this fleet, select the radio box(es) next to each device and enter the number of trucks equipped with the device by <u>engine</u> (not vehicle) model year.

4. When done, select **VALIDATE SCREEN** to make sure you have filled out everything properly on this screen.

5. Select the **HOME** button to return to the Home screen.

Next Steps

*** If you are using PM reduction equipment on model year 2006 or earlier trucks,
select the PM Reduction tab at the top of the screen to proceed to this section.
Instructions for data entry are provided above this box. *******

If you have finished entering data for this fuel type, select the tab for the next fuel type and complete all screens as indicated above.

If you have finished inputting data for all of your fuel types, select the  button to return to the Home screen.

REMEMBER:

*** If additional fuel types are represented in this fleet,
you must complete the Engine Model Year & Class and Activity Information screens for each of the fuel types you operate.

Select the tab for the next fuel type and complete the *Engine Model Year & Class and Activity Information screens* as indicated above—the steps will be the same or similar to those for the Diesel Tab. *Do NOT move on if you have not completed the data input for each of your fuel types.****

Once finished entering data for all of your fuel types, select the  button to return to the Home screen and follow the instructions to submit your tool to EPA.

Validating Your Data

The Truck Carrier Tool has validation checks embedded at different points in the Tool to ensure data quality. For example, checks on total mileage or number of vehicles are made before exiting the Payload and Volume Calculators. However, additional validation checks are made before exiting other data entry screens throughout the Tool. Critical checks are made regarding calculated miles per gallon (by class), annual miles per truck, and a number of other inputs, to help ensure the reasonableness and quality of Partner data.

In most instances the validation ranges are based on Truck Partner data distributions from the 2011 calendar year. (Refer to the **Truck Carrier Tool Technical Documentation** (found on the website at http://epa.gov/smartway/partnership/trucks.htm) for a detailed discussion of all validation criteria used within the Tool.)

At any time, you can select the [   ] button at the bottom of the screens to make sure you have filled out everything properly before proceeding to the next screen.

Any time you leave a particular screen, the Tool will automatically perform a *screen* validation.

Any time you return to the Home screen, the Tool will automatically perform a *fleet-level* validation.

If you select the [   ] button and no potential errors are identified, the following message will be displayed. Select **OK** to proceed.

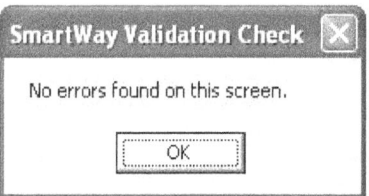

Figure 50: Validation Check Notification – No Errors

If missing or potentially erroneous inputs are identified, you will see the following message (**Figure 51**).

Select **Yes** to see a complete list of validation errors/warnings. An example **Validation Check** results screen is provided in **Figure 52**.

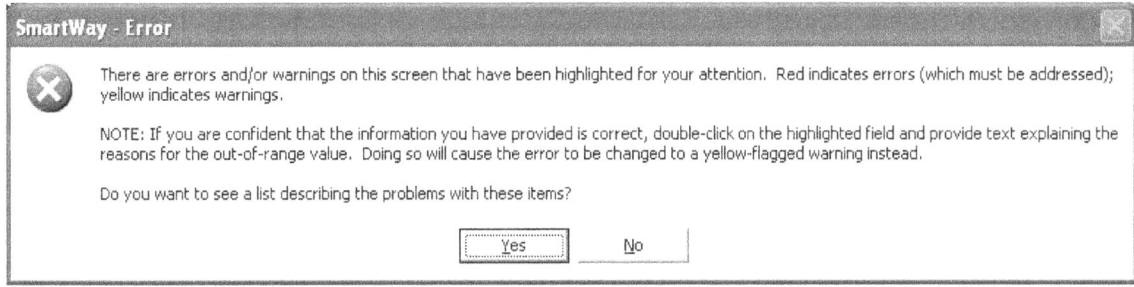

SmartWay - Error

There are errors and/or warnings on this screen that have been highlighted for your attention. Red indicates errors (which must be addressed); yellow indicates warnings.

NOTE: If you are confident that the information you have provided is correct, double-click on the highlighted field and provide text explaining the reasons for the out-of-range value. Doing so will cause the error to be changed to a yellow-flagged warning instead.

Do you want to see a list describing the problems with these items?

Yes   No

Figure 51: Validation Check Notification – with Errors/Warnings

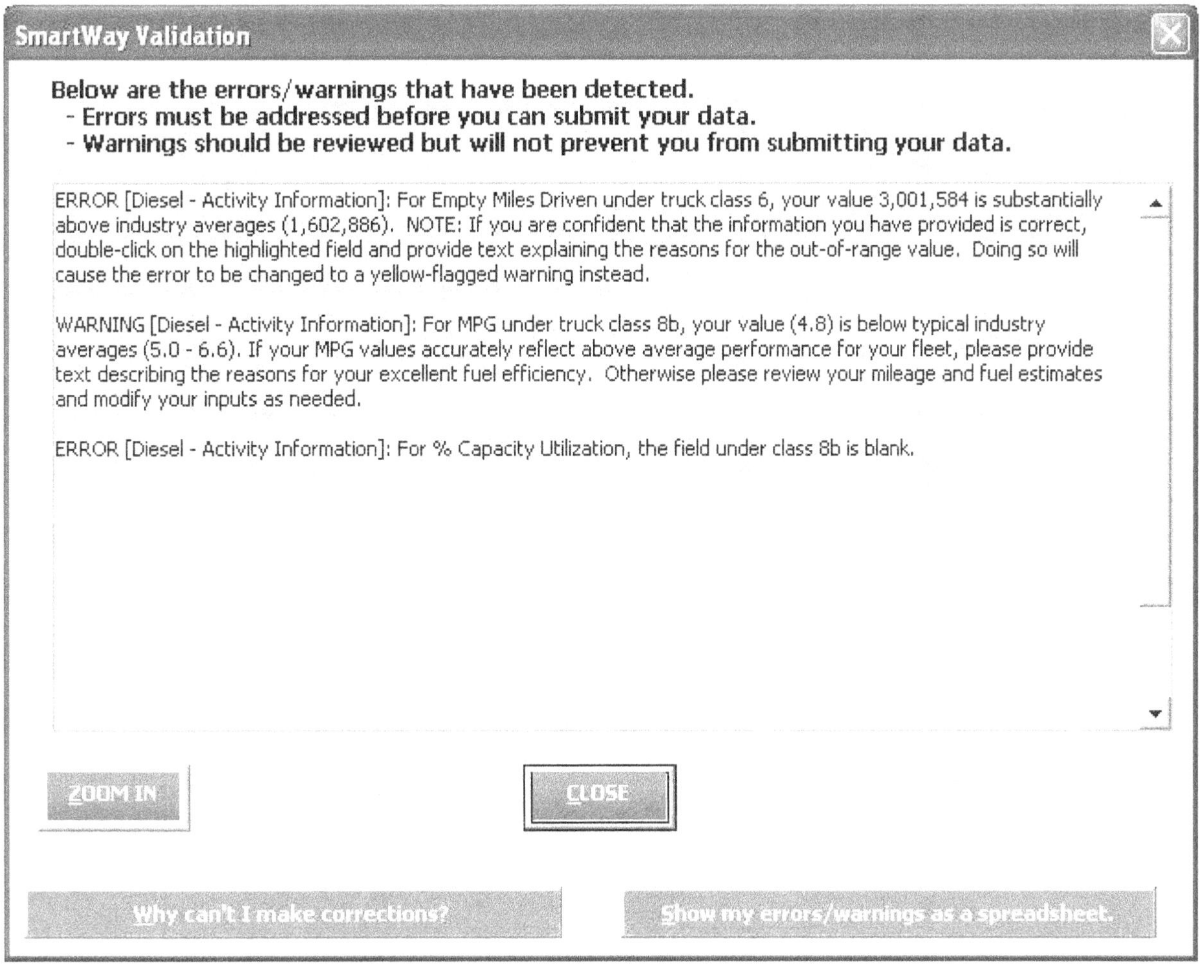

SmartWay Validation

Below are the errors/warnings that have been detected.
 - Errors must be addressed before you can submit your data.
 - Warnings should be reviewed but will not prevent you from submitting your data.

ERROR [Diesel - Activity Information]: For Empty Miles Driven under truck class 6, your value 3,001,584 is substantially above industry averages (1,602,886). NOTE: If you are confident that the information you have provided is correct, double-click on the highlighted field and provide text explaining the reasons for the out-of-range value. Doing so will cause the error to be changed to a yellow-flagged warning instead.

WARNING [Diesel - Activity Information]: For MPG under truck class 8b, your value (4.8) is below typical industry averages (5.0 - 6.6). If your MPG values accurately reflect above average performance for your fleet, please provide text describing the reasons for your excellent fuel efficiency. Otherwise please review your mileage and fuel estimates and modify your inputs as needed.

ERROR [Diesel - Activity Information]: For % Capacity Utilization, the field under class 8b is blank.

ZOOM IN   CLOSE

Why can't I make corrections?   Show my errors/warnings as a spreadsheet.

Figure 52: Validation Check Screen – Example Results

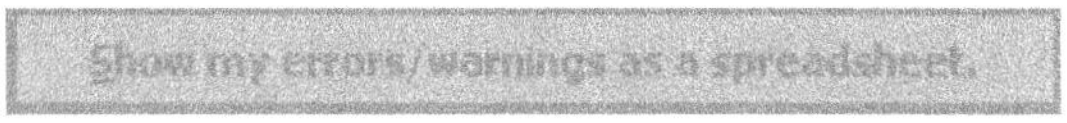

The [Why can't I make corrections?] and [Show my errors/warnings as a spreadsheet.] buttons at the bottom of this screen provide additional information to assist you in correcting any errors. Specific information is also provided regarding any values that are unusually high or low compared to industry averages (e.g., see the second entry). In this case, three possible data entry problems were identified for correction: one involving a suspiciously high "average empty miles per year" value, one for a low MPG value, and one for a missing value. Two of the three messages involve errors that <u>must be addressed</u> through modification of the data entry (message #3), or by providing a text justification of the value by double-clicking on the highlighted cell (message #1). Both of these items must be addressed before the Tool may be submitted to EPA. The remaining item (#2) is listed as a warning, and does not have to be addressed before submittal to EPA, although you may provide documentation for this value by double-clicking on this cell if you wish.

Once you close the Validation check screen, you will see warnings and errors highlighted in a red or yellow on the Activity Information screen, as shown in **Figure 53**. Entering new data in a highlighted cell will cause the colored shading to disappear, ***even if the new value is still invalid***. To check the validity of your newly entered data, select the ⬚⬚⬚⬚⬚ button again and repeat the procedure described above as necessary.

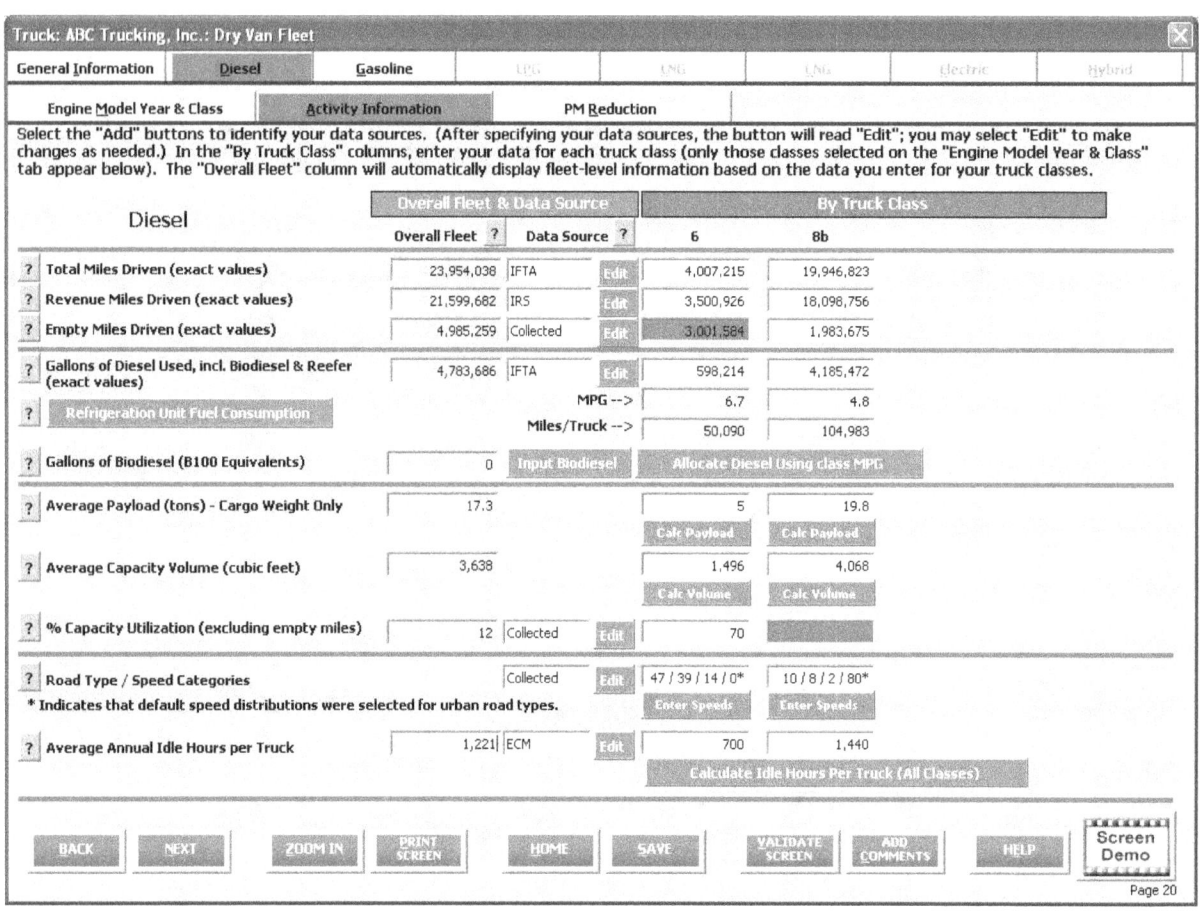

Figure 53: Validation Check – Highlighted Activity Screen

Follow these steps to check the different types of validation warnings/errors.

- First ensure that any highlighted blank cells are completed and that all standard range checks are satisfied (e.g., all percentages must be between 0 and 100, etc.).

- If "Total Miles Driven" entries are highlighted for any truck class, this means that the expected annual miles per truck have been exceeded. Calculated annual mileage values are shown on the Activity Information screen below the calculated MPG values. You may revise these values by adjusting either "Total Miles Driven" OR the "Total Trucks" (on the Engine Model Year & Class screen). Unless the calculated mileage value is greater than the absolute maximum allowed, you may leave these values as they are, and simply enter a text explanation (for "red" warning values). In this case your fleet will be flagged "Complete with Warnings" on the Home screen.

- If "Revenue Miles" or "Empty Miles" are highlighted, check their values relative to "Total Miles Driven" and to one another. For example, explanations are generally required if "Revenue Miles" are roughly less than 50% of "Total Miles Driven," or if "Empty Miles" are greater than approximately 50% of "Total Miles Driven", although specific validation cutoffs vary depending upon truck class, body type and operational category.

- "% Capacity Utilization" values of less than ~40-70% must be explained.

- "Average Idle Hours per Truck" outside typical partner values must be explained.

- If any calculated MPG values are highlighted, you may revise these values by changing either "Total Miles Driven" or the number of gallons of the fuel type used. You may also double-click on the highlighted MPG cell to obtain a popup screen with additional information, as shown in **Figure 54**. This form presents the expected MPG range based on typical SmartWay Partner performance, as well as the calculated value for your truck class, and the "Severity" of the error identified. (Level 1 Severity indicates the value is significantly different from partner averages, while Level 2 Severity indicates a moderately out of range value.) If you believe this value is accurate, enter a brief summary in the text box explaining the reasons for the high/low MPG value and select **OK** to return to the Activity Information screen. (In this case your fleet will be flagged "Complete with Warnings" on the Home screen.) Otherwise select **CANCEL** to revise your total miles and/or gallons used.

- If payload, volume, and/or idle hour values are highlighted on the **Activity** screen, warnings and/or explanations can be viewed by opening the appropriate calculator(s).

- Possible rounding errors for miles and/or gallons are not highlighted on the **Activity** screen but are described in the detailed Validation list popup screen.

SmartWay Range Validation Form

Data you provided is outside the expected range. If you are confident that the information you entered is correct, please provide supporting details in the space below. Otherwise, select Cancel to return to the form, so you may re-enter a new value for this field.

Field Name: Class 8b mpg

Out of Range Message: WARNING [Diesel - Activity Information]: For MPG under truck class 8b, your value (4.8) is below typical industry averages (5.0 - 6.6). If your MPG values accurately reflect above average performance for your fleet, please provide text describing the reasons for your excellent fuel efficiency. Otherwise please review your mileage and fuel estimates and modify your inputs as needed.

Value You Provided: 4.8

Severity of Error: Level 2*

Explanation:

* Level 1 indicates that the value is significantly out of expected range. Level 2 indicates that the value is moderately out of expected range.

[ZOOM IN] [OK] [CANCEL]

Figure 54: Validation Check – Example MPG Range Validation Form

Other checks are made regarding absolute minimum and maximum inputs for miles per truck, MPG, payload, volume, and idle hours, as well as checks regarding implicit cargo density based on payload, volume, and capacity utilization inputs. When absolute minimum or maximum values are exceeded, the Partner must revise the associated input values before proceeding (i.e., these values cannot be "explained away").

The Truck Carrier Tool also utilizes a number of "logical" validation checks to ensure the reasonableness of various inputs and category selections. For example, checks are applied regarding the consistency between body types specified in the **Fleet Characterization** section and those selected in the payload and volume calculators. If inconsistencies are found, in most instances a simple warning is indicated, and text explanations are not required. For a complete listing of all validation checks used in the Tool, refer to the Truck Carrier Tool Technical Documentation.

Once all potential validation errors have been corrected or otherwise verified, select [   ] to proceed to the next step.

Once you have returned to the Home screen, notice that the fleet that you filled out and validated now identifies its status as "Complete." You may now highlight the next fleet if you have another one to complete. In the example below, the fleet listed as "Example Dray Fleet" remains to be finished. Fill out unfinished fleets in the same manner as the previous fleet.

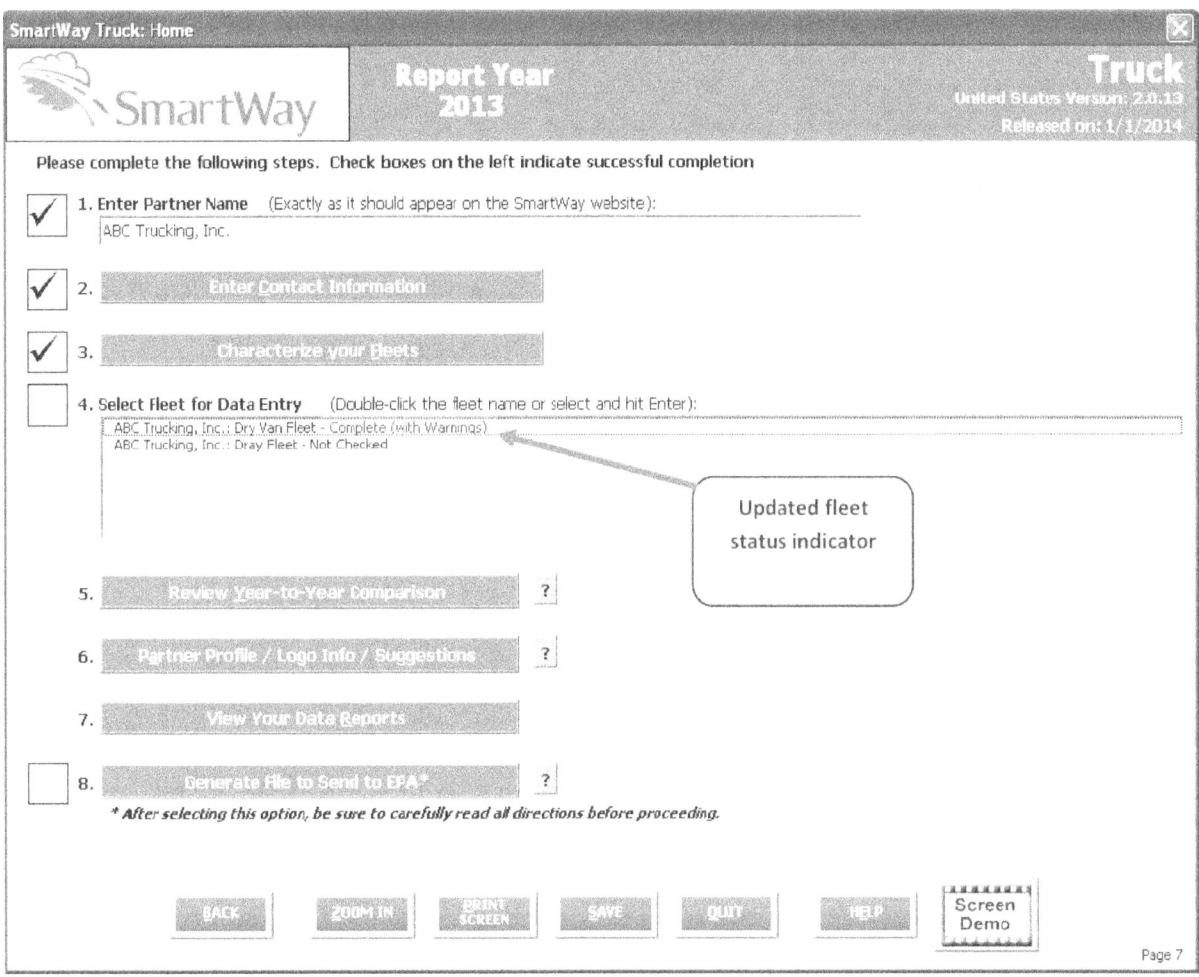

Figure 55: Home Screen - After Completing the Data Entry/Modification Process for First Fleet

Once you have filled out information for all your fleets, be sure that all fleets show "Complete" or "Complete with Warnings" beside the name. If, and only if, all fleets are marked as such, you can move on to the next step. If one or more fleets are not marked as such, review the data you entered for errors or omissions.

Viewing Year-to-Year Comparison

The **Year-to-Year Comparison Report** allows the user to compare the fleet characteristics and activity values, as well as CO_2 performance metrics for the current reporting year with the previous year. This report is particularly helpful in identifying any changes that may have occurred since your last reporting period, determining trends in activity and performance over multiple years, and performing general quality assurance of the inputs used for your current Tool. You can access this report by selecting

Review Year-to-Year Comparison under item #5 on the Home screen.

Figure 56 shows the data entry screen for the comparison report. Note that your most recent year's data is already "loaded," including the data you have entered for your current fleet(s). In order to load data for the previous year fleets, select the button. A screen will appear allowing you to specify the location of your prior year file, using the button as shown in **Figure 57**.

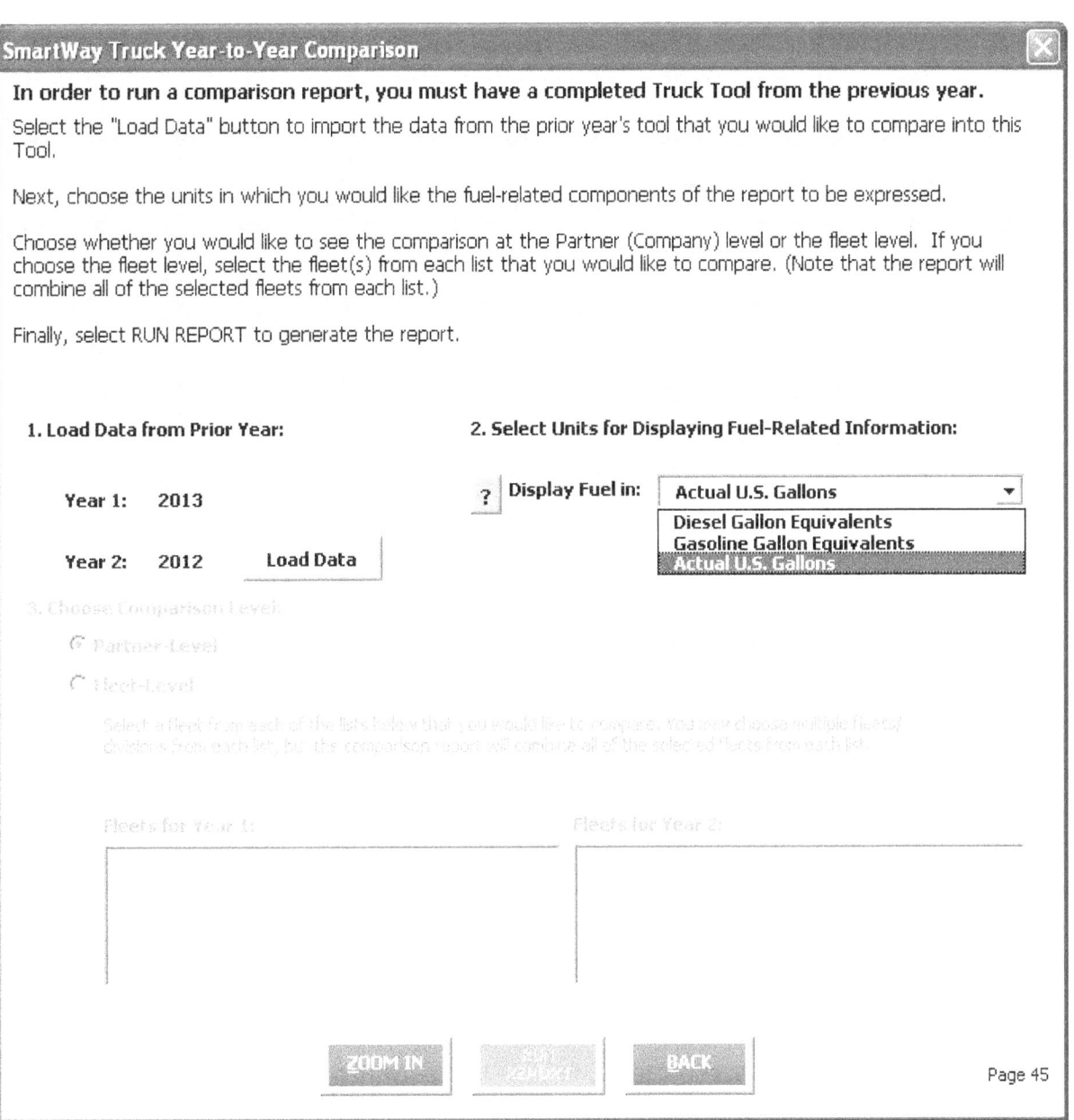

Figure 56: Year-to-Year Comparison Report Input Screen

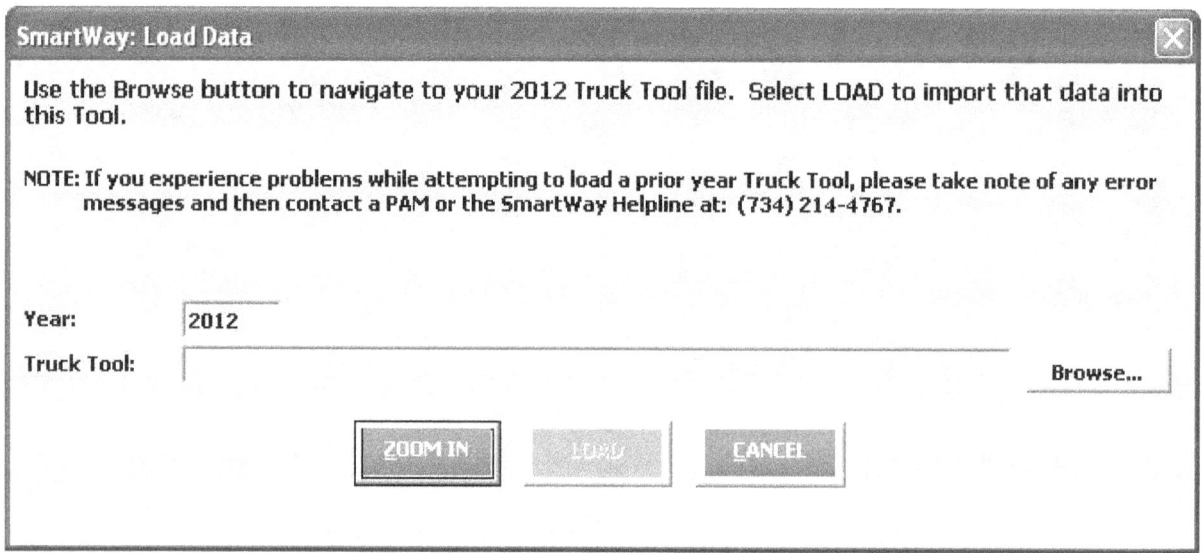

Figure 57: Data Load Screen for Year-to-Year Comparison Report

Once your prior year data loaded into the Truck Carrier Tool, you should specify how to display your fuel information: in diesel gallon equivalents, gasoline gallon equivalents, or simply in actual U.S. gallons. Specifying diesel or gasoline gallon equivalents is useful for comparing miles per gallon metrics when a fleet contains multiple fuel types.

Next you can specify whether you want to compare one or more individual fleets (select "Fleet-Level") or if instead the comparison should aggregate all fleets (select "Company-Level"). Once specified, select the "RUN REPORT" button to view the line-item data entry comparisons. An example report is presented in **Figure 58.**

	Change %	Change Amount	2011	2010
Fleet/Division Name(s)			ABC Trucking: North America	ABC Trucking: North America
Primary Contact(s)			Cookie Monster	Cookie Monster
Contact Phone(s)			(555) 555-5555	(555) 555-5555
SCAC(s)			COOK	COOK
MCN(s)			#555555	#555555
Fleet Type(s)			For-Hire	For-Hire
Operation / Body Type			Drayage / Mixed	Drayage / Mixed
Operation Type			TL(10%) / LTL(10%) / Drayage(80%)	TL(10%) / LTL(10%) / Drayage(80%)
Body Type			DryVan(35%) / Reefer(15%) / Chassis(50%)	DryVan(35%) / Reefer(15%) / Chassis(50%)
Commodities		X	Cereal Grains; Milled Grain and Bakery Products; Other Prepared Foodstuffs; Refrigerated Food and Frozen Food; Mixed Freight	Cereal Grains; Milled Grain and Bakery Products; Other Prepared Foodstuffs; Refrigerated Food and Frozen Food
Average Cube Out	-7.0%	-6%	80%	86%
Miles Per Gallon (DGE)*	4.8%	0.27	5.98	5.71
Miles Per Truck	-14.2%	-14,351	86,933	101,285
Average Payload (tons)	-4.9%	-0.9	17.1	18.0
Average Volume (cu ft)	-4.9%	-119	2,310	2,429
Average Capacity Utilization	-1.5%	-1%	88%	89%
Percent Highway Driving			74%	74%
Percent Urban Driving (0-25)	-4.1%	-1%	12%	13%
Percent Urban Driving (25-50)			11%	11%
Percent Urban Driving (50+)	22.3%	1%	4%	3%
Average Idle Hours Per Truck	-0.4%	-5	1,191	1,196
Total Trucks	4.2%	3	75	72
Class 2b	0.0%	0	2	2
Class 3	100.0%	5	10	5
Class 4	0.0%	0	4	4
Class 5	50.0%	2	6	4
Class 6			-	-
Class 7			-	-
Class 8a			-	-
Class 8b	-7.0%	-4	53	57
Total Miles Driven	-10.6%	-772,500	6,520,000	7,292,500
Revenue Miles Driven	-13.4%	-854,500	5,515,000	6,369,500
Empty Miles Driven	8.9%	82,000	1,005,000	923,000
Total Gallons of Fuel (DGE)*	-14.7%	-187,869	1,090,236	1,278,104
Diesel (DGE)*	-15.3%	-186,668	1,033,332	1,220,000
Biodiesel (DGE)*			-	-
Gasoline (DGE)*	30.1%	2,504	10,830	8,327
Ethanol (DGE)*	30.1%	140	604	464
Other (DGE)*	-7.8%	-3,844	45,470	49,314
Tons of CO_2	-14.7%	-2,097	12,127	14,224
CO_2 g/mile	-4.6%	-82	1,687	1,769
CO_2 g/ton-mile			99	98
CO_2 g/1000 cuft-mile	0.3%	2	730	729
CO_2 g/utilized 1000 cuft-mile	2.3%	19	830	811

* Diesel Gallon Equivalents

Figure 58: Example Year-to-Year Comparison Report

The example above indicates that a change in commodities has occurred over a year, with no other changes in basic contact and fleet characteristic information. Comparisons are also shown for a number of activity parameters including:

- average payload
- average volume
- average capacity utilization
- percent driving by road type and speed bin
- average idle hours per truck
- number of trucks by class
- total, revenue, and empty miles driven
- total gallons of fuel used (in this case, in diesel gallon equivalents (DGE))

A variety of calculated annual performance metrics are also shown, including:

- miles per gallon (here, in diesel gallon equivalents)
- miles per truck
- tons of CO_2
- grams CO_2 per mile
- grams CO_2 per ton-mile
- grams CO_2 per 1,000 cubic foot-miles
- grams CO_2 per utilized 1,000 cubic foot-miles

Percentage changes are also indicated for each of these items, relative to the earlier year baseline values.

Providing Additional Information

After completing Steps 1 – 4 on the Home screen, you may provide EPA with additional information regarding your company, potential use of the SmartWay Logo, and general feedback regarding the SmartWay program. This information is optional and is not required in order to submit your Truck Carrier Tool data to EPA. Selecting the [Partner Profile / Logo Info / Suggestions] button on the Home screen will open the Partner Information form (see **Figure 59**).

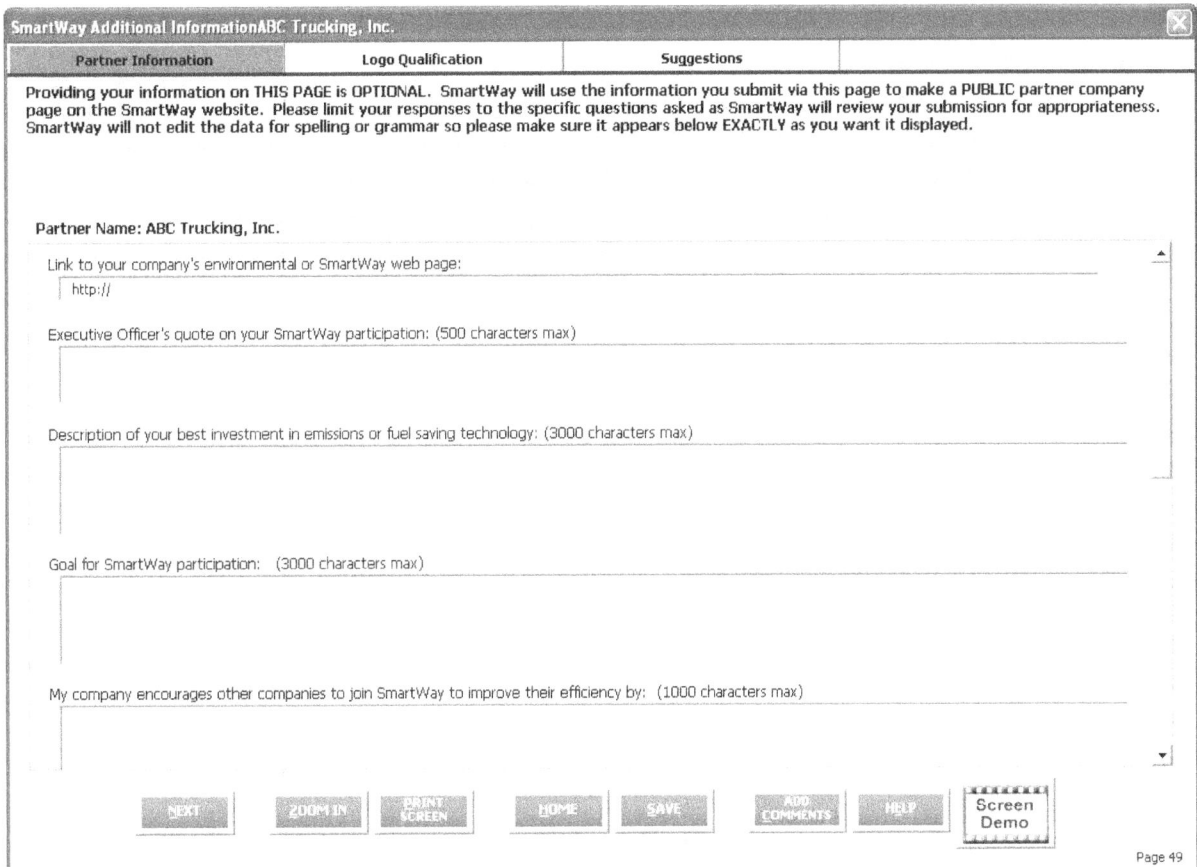

Figure 59: Partner Information Screen

Use this page to provide information you would like to share publicly. Information should be entered in the text boxes displayed. Use the scroll bar to the right to display additional questions. SmartWay will use this information to create a SmartWay Partner profile page for your company on the SmartWay website. You do not need to complete every question. SmartWay will NOT edit for spelling or grammar, so make sure the text is exactly as you wish it to appear. If your company contains public relations functions, you may want them to review this information before submittal, however, keep your SmartWay due date in mind. SmartWay WILL review this data for appropriate content. Information provided should be informational in nature, and speak to the question.

Figure 603 shows the Logo Qualification screen. The SmartWay Partner Logo is provided at the company level to Partners in good standing in the SmartWay program. To be in good standing you must submit your SmartWay materials by the appropriate due date. Due dates are posted on the EPA website at epa.gov/smartway.. Tractor and Trailer Logos require the approval of the SmartWay Brand Manager (Jackson-Stephens.joann@epa.gov). Tractor and trailer logos may be placed directly on your equipment to display your fleet's qualification. Use this screen to provide information regarding your qualification for, and planned use of, the Partner, Truck and Trailer Logos.   Use the scroll bars to display more questions and input fields. Additional information on Logo qualification and use can be found by

selecting the SmartWay Partner Logo Information button or the View the SmartWay Designated Tractor specifications and View the SmartWay Designated Trailer specifications links.

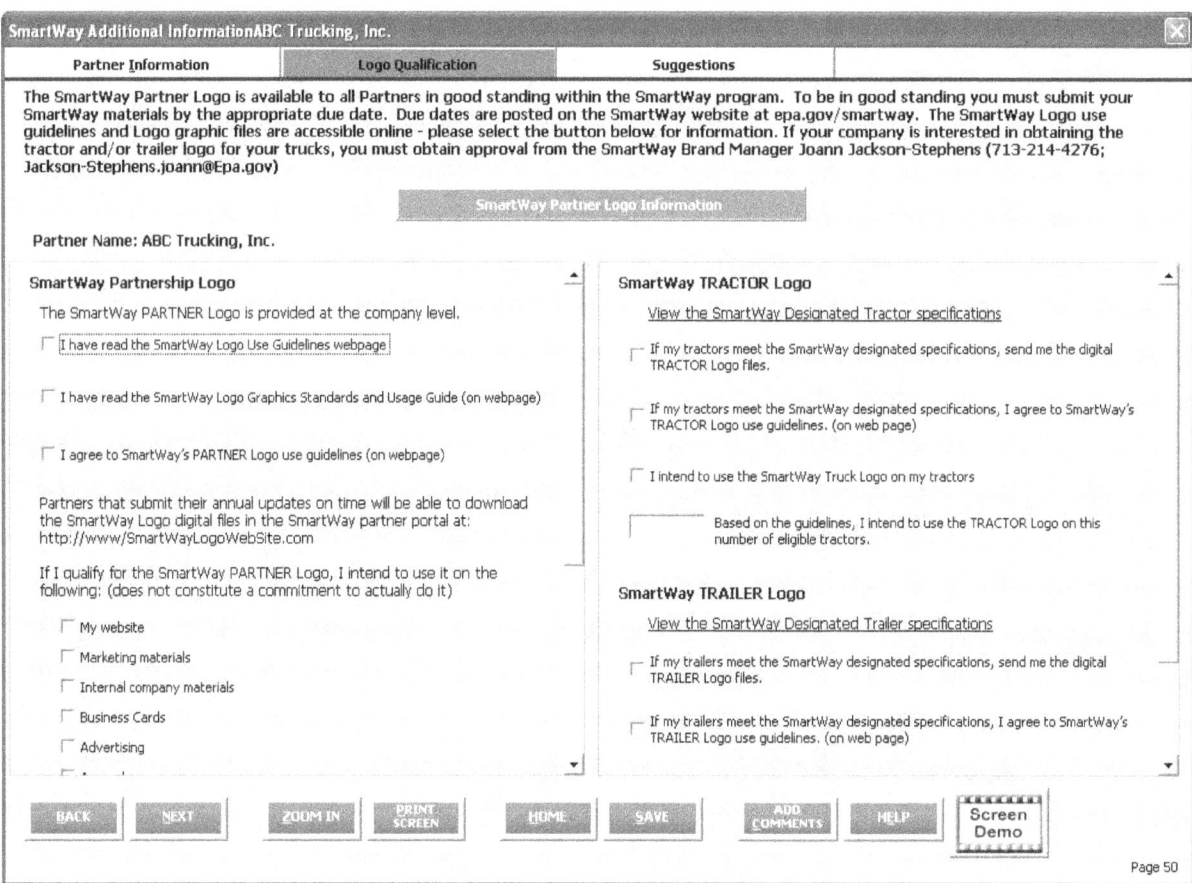

Figure 60: Logo Qualification Screen

Figure 61 displays the Suggestions screen. You may respond to some or all of these questions in order to provide feedback regarding the various aspects of the SmartWay program. SmartWay values your feedback. Any information you provide will be used to improve the program, and will be kept confidential. Use the scroll bar to the right to display additional questions.

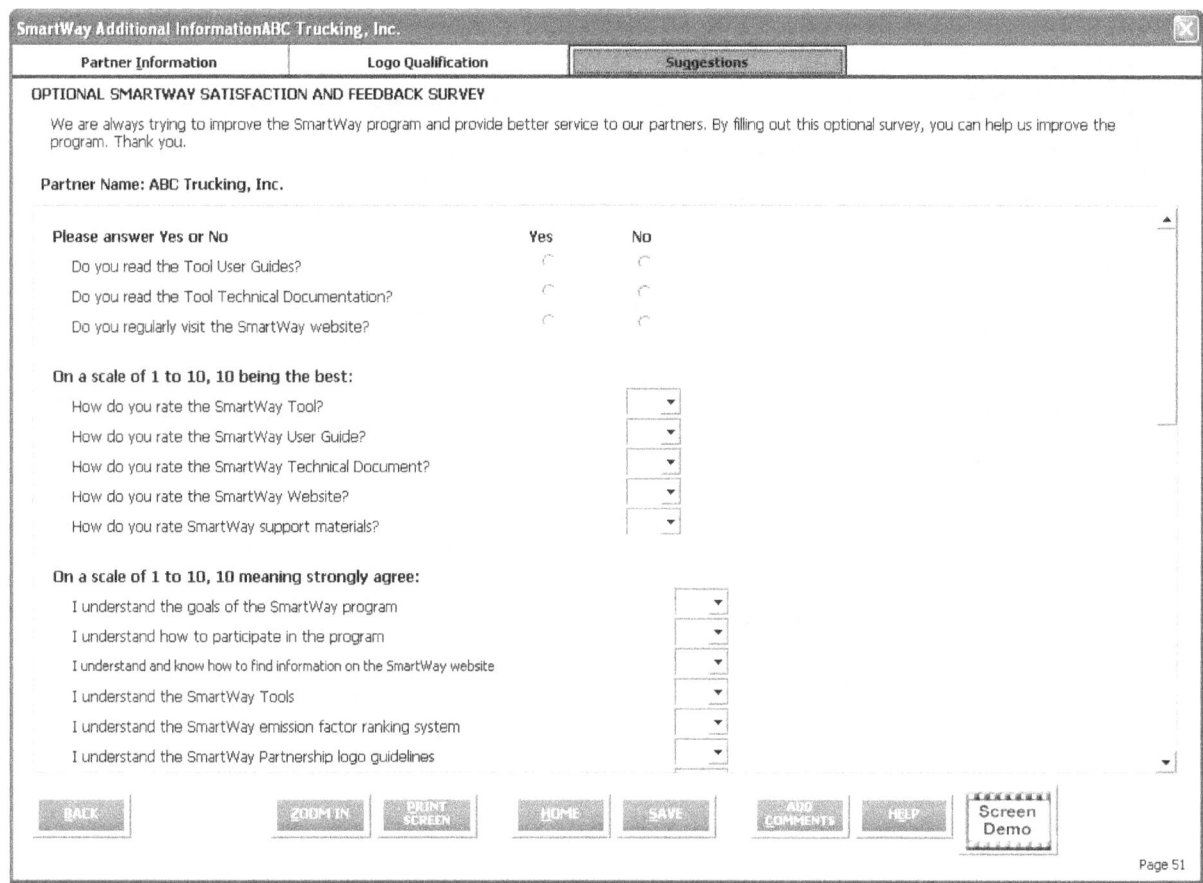

Figure 61: Suggestions Screen

Once you have completed these three screens, select the [   ] button to return to the Home screen.

Viewing Reports

Once you are ready to continue, select [View Your Data Reports] to go to the following screen:

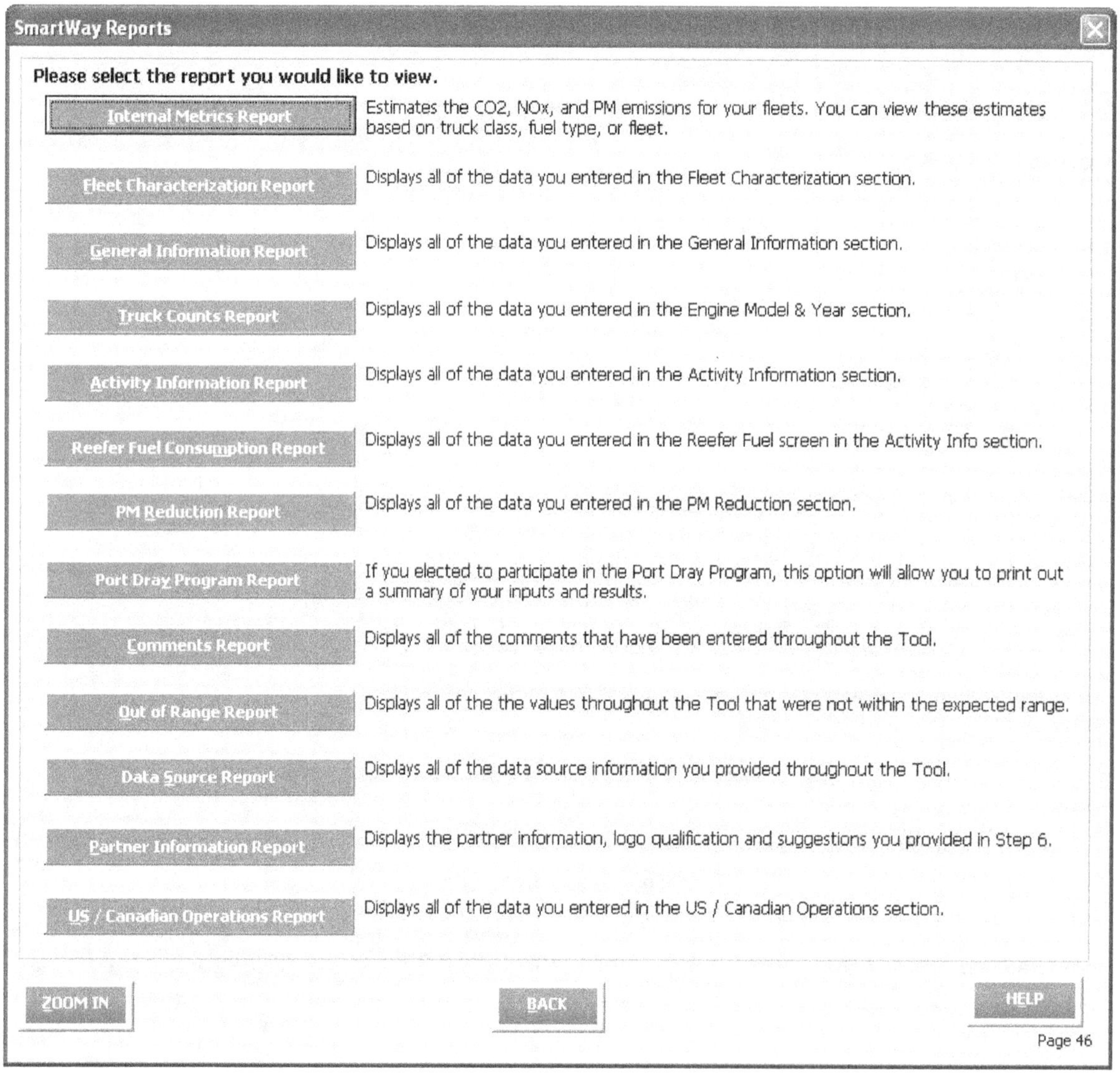

SmartWay Reports

Please select the report you would like to view.

Internal Metrics Report	Estimates the CO2, NOx, and PM emissions for your fleets. You can view these estimates based on truck class, fuel type, or fleet.
Fleet Characterization Report	Displays all of the data you entered in the Fleet Characterization section.
General Information Report	Displays all of the data you entered in the General Information section.
Truck Counts Report	Displays all of the data you entered in the Engine Model & Year section.
Activity Information Report	Displays all of the data you entered in the Activity Information section.
Reefer Fuel Consumption Report	Displays all of the data you entered in the Reefer Fuel screen in the Activity Info section.
PM Reduction Report	Displays all of the data you entered in the PM Reduction section.
Port Dray Program Report	If you elected to participate in the Port Dray Program, this option will allow you to print out a summary of your inputs and results.
Comments Report	Displays all of the comments that have been entered throughout the Tool.
Out of Range Report	Displays all of the the values throughout the Tool that were not within the expected range.
Data Source Report	Displays all of the data source information you provided throughout the Tool.
Partner Information Report	Displays the partner information, logo qualification and suggestions you provided in Step 6.
US / Canadian Operations Report	Displays all of the data you entered in the US / Canadian Operations section.

[ZOOM IN]   [BACK]   [HELP]

Page 46

Figure 62: View Reports Menu

Selecting any of the green buttons on this screen will display the indicated data. Several reports summarize the data you entered on specific data input screens, such as the **Truck Counts Report** (corresponding to the Engine Model Year & Class screen) and the **Activity Information Report** (corresponding to the Activity Information screen). The **Out of Range Report** may be particularly useful in identifying and addressing those inputs flagged as potentially erroneous during fleet validation.

Select the **Internal Metrics Report** button to review the performance of your fleets in terms of grams per mile and grams per ton-mile, by pollutant type. After selecting this button you will be asked to specify the level of detail/aggregation you wish to display as shown in **Figure 63**.

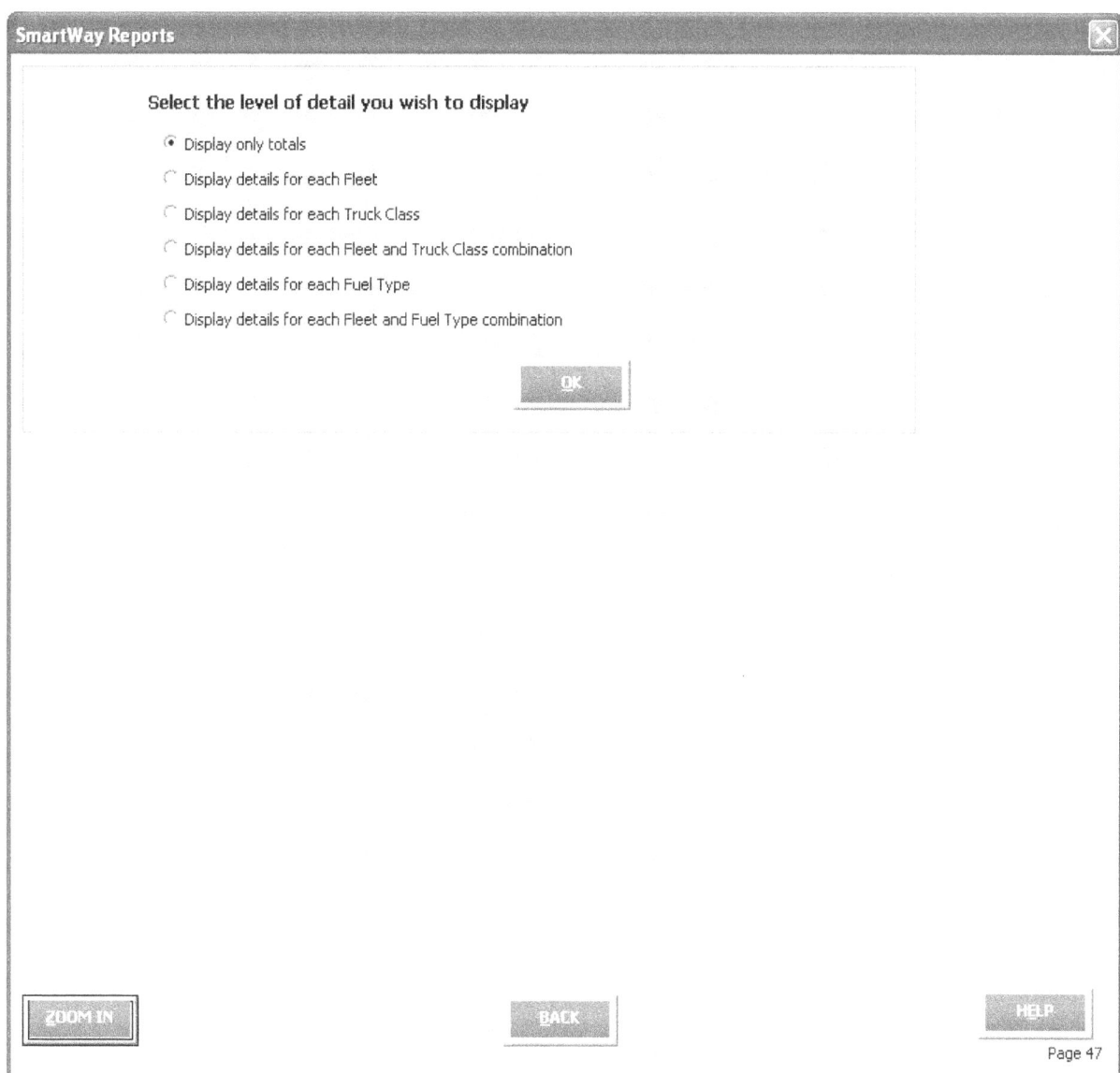

Figure 63: Internal Metrics Report Selection Menu

Selecting any of the summary report types will take you to a screen allowing you to preview and print your reports. **Figure 64** shows one of many sample reports that the Tool can generate for you; in this case summarizing emission reductions in tons and assorted performance metrics at the Partner Total level (including all fleets). These reports will prove useful for your company's evaluation and improvement efforts.

SmartWay

CO2

Total Tons (Short)	Total Miles				Loaded Miles				Revenue Miles			
	Grams per Mile (Total)	Grams per Average Payload Ton-Mile (Total)	Grams per Thousand Cubic Foot-Miles (Total)	Grams per Thousand Utilized Cubic Foot-Miles (Total)	Grams per Mile (Loaded)	Grams per Average Payload Ton-Mile (Loaded)	Grams per Thousand Cubic Foot-Miles (Loaded)	Grams per Thousand Utilized Cubic Foot-Miles (Loaded)	Grams per Mile (Revenue)	Grams per Average Payload Ton-Mile (Revenue)	Grams per Thousand Cubic Foot-Miles (Revenue)	Grams per Thousand Utilized Cubic Foot-Miles (Revenue)
Partner Total 53650	2.033	117	896	690	2.567	135	704	601	2.565	130	661	78

NOx

Total Tons (Short)	Total Miles				Loaded Miles				Revenue Miles			
	Grams per Mile (Total)	Grams per Average Payload Ton-Mile (Total)	Grams per Thousand Cubic Foot-Miles (Total)	Grams per Thousand Utilized Cubic Foot-Miles (Total)	Grams per Mile (Loaded)	Grams per Average Payload Ton-Mile (Loaded)	Grams per Thousand Cubic Foot-Miles (Loaded)	Grams per Thousand Utilized Cubic Foot-Miles (Loaded)	Grams per Mile (Revenue)	Grams per Average Payload Ton-Mile (Revenue)	Grams per Thousand Cubic Foot-Miles (Revenue)	Grams per Thousand Utilized Cubic Foot-Miles (Revenue)
Partner Total 197.1	7.5	0.43	3.190	2.536	9.4	0.50	2.570	2.342	9.3	0.48	2.429	2.30

PM2.5

Total Tons (Short)	Total Miles				Loaded Miles				Revenue Miles			
	Grams per Mile (Total)	Grams per Average Payload Ton-Mile (Total)	Grams per Thousand Cubic Foot-Miles (Total)	Grams per Thousand Utilized Cubic Foot-Miles (Total)	Grams per Mile (Loaded)	Grams per Average Payload Ton-Mile (Loaded)	Grams per Thousand Cubic Foot-Miles (Loaded)	Grams per Thousand Utilized Cubic Foot-Miles (Loaded)	Grams per Mile (Revenue)	Grams per Average Payload Ton-Mile (Revenue)	Grams per Thousand Cubic Foot-Miles (Revenue)	Grams per Thousand Utilized Cubic Foot-Miles (Revenue)
Partner Total 9.55	0.362	0.0209	0.1065	0.1228	0.457	0.0240	0.1248	0.1425	0.401	0.0231	0.1117	0.135

PM10

Total Tons (Short)	Total Miles				Loaded Miles				Revenue Miles			
	Grams per Mile (Total)	Grams per Average Payload Ton-Mile (Total)	Grams per Thousand Cubic Foot-Miles (Total)	Grams per Thousand Utilized Cubic Foot-Miles (Total)	Grams per Mile (Loaded)	Grams per Average Payload Ton-Mile (Loaded)	Grams per Thousand Cubic Foot-Miles (Loaded)	Grams per Thousand Utilized Cubic Foot-Miles (Loaded)	Grams per Mile (Revenue)	Grams per Average Payload Ton-Mile (Revenue)	Grams per Thousand Cubic Foot-Miles (Revenue)	Grams per Thousand Utilized Cubic Foot-Miles (Revenue)
Partner Total 9.85	0.373	0.0215	0.1098	0.1266	0.471	0.0248	0.1287	0.1469	0.414	0.0238	0.1213	0.136

Figure 64: Example of Completed Report

Submitting Data to SmartWay

Congratulations! You are now ready to send your data to EPA.

Select the [Generate File to Send to EPA*] button, which will open the following screen.

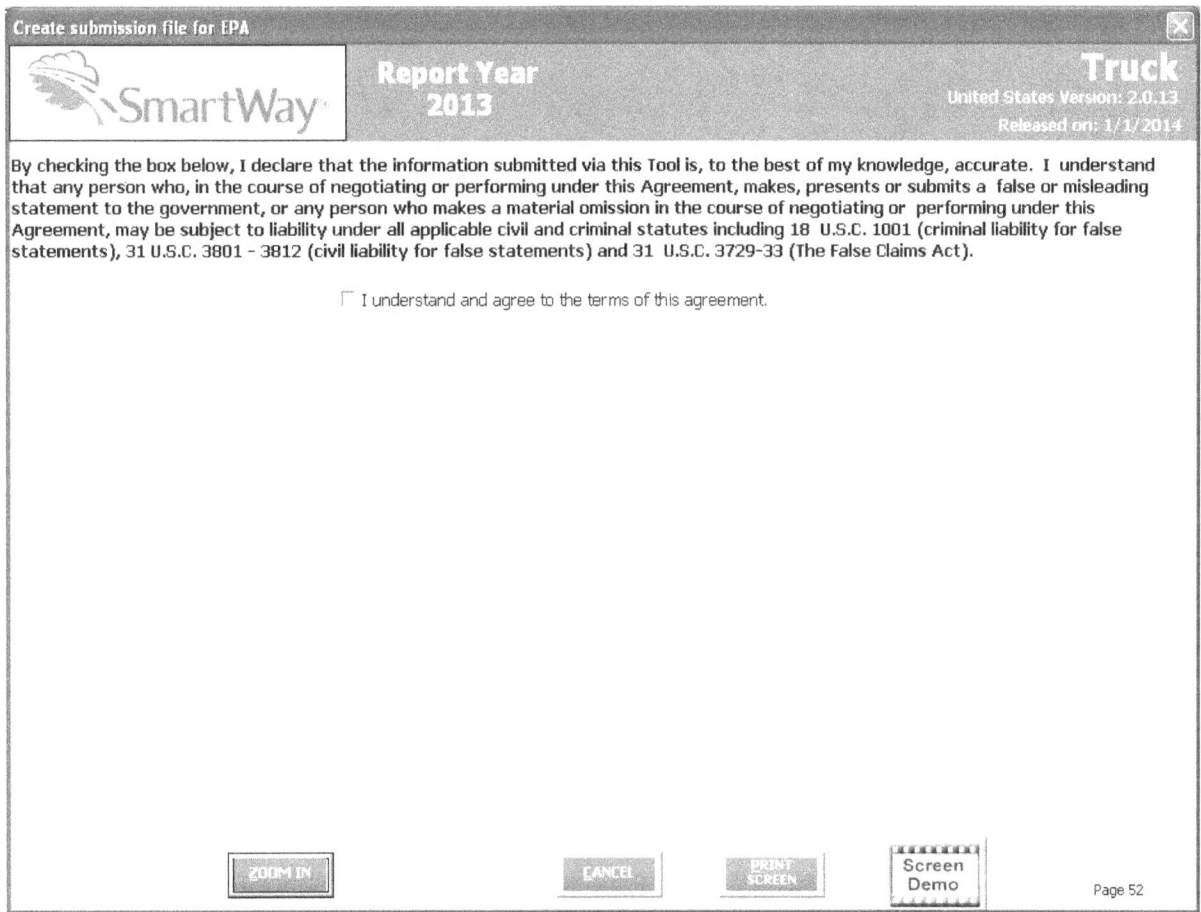

Figure 65: Creating Submission File for EPA

Select the checkbox to indicate you understand the terms of the SmartWay Partnership Agreement once again. Next, a question will appear asking if you are an existing SmartWay Partner (Y/N). If you are, a question will appear asking if you submitted your data the previous reporting year. If so, you must then enter your Annual Submission ID, which has been sent to you by your SmartWay PAM via email. If you cannot locate your submission ID you can select the "Email me my SmartWay ID" button to have your ID sent to you.

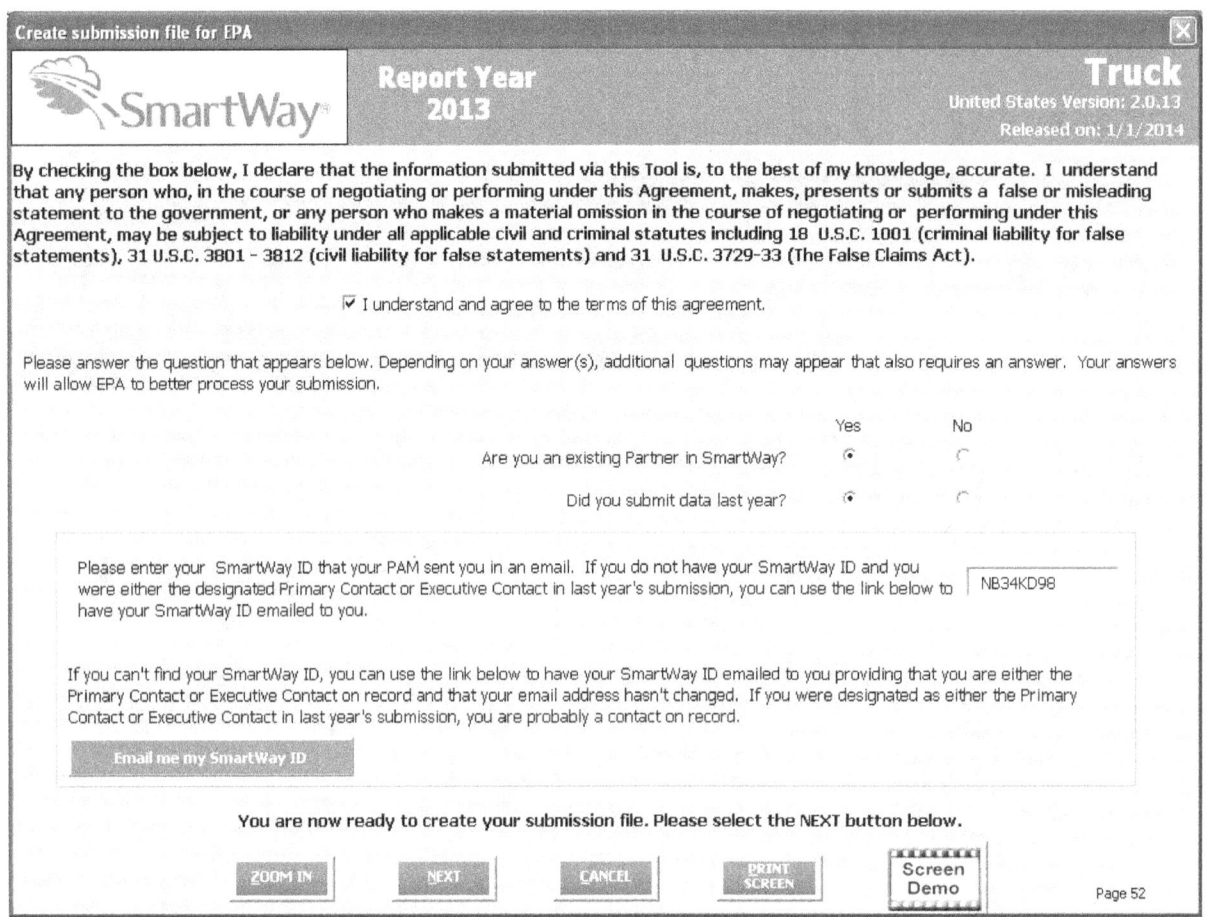

Figure 66: Completed Submission File Screen

When ready, select **NEXT** to create a file with the following naming convention:

Truck_PartnerName_ Year_ V0.xml [12]

For example, Truck_ABCompany_ 2013 _V0.xml

where **PartnerName** is your company's name as entered for Step 1 on the Home screen, and **Year** indicates the year for which you are submitting your data.

Next specify the folder where you would like to save the. xml file, and the following screen will appear.

---

[12] If you create the XML file multiple times the file name will increment each time (e.g., Truck_ABCompany_2013_V1.XML for the second iteration, etc.

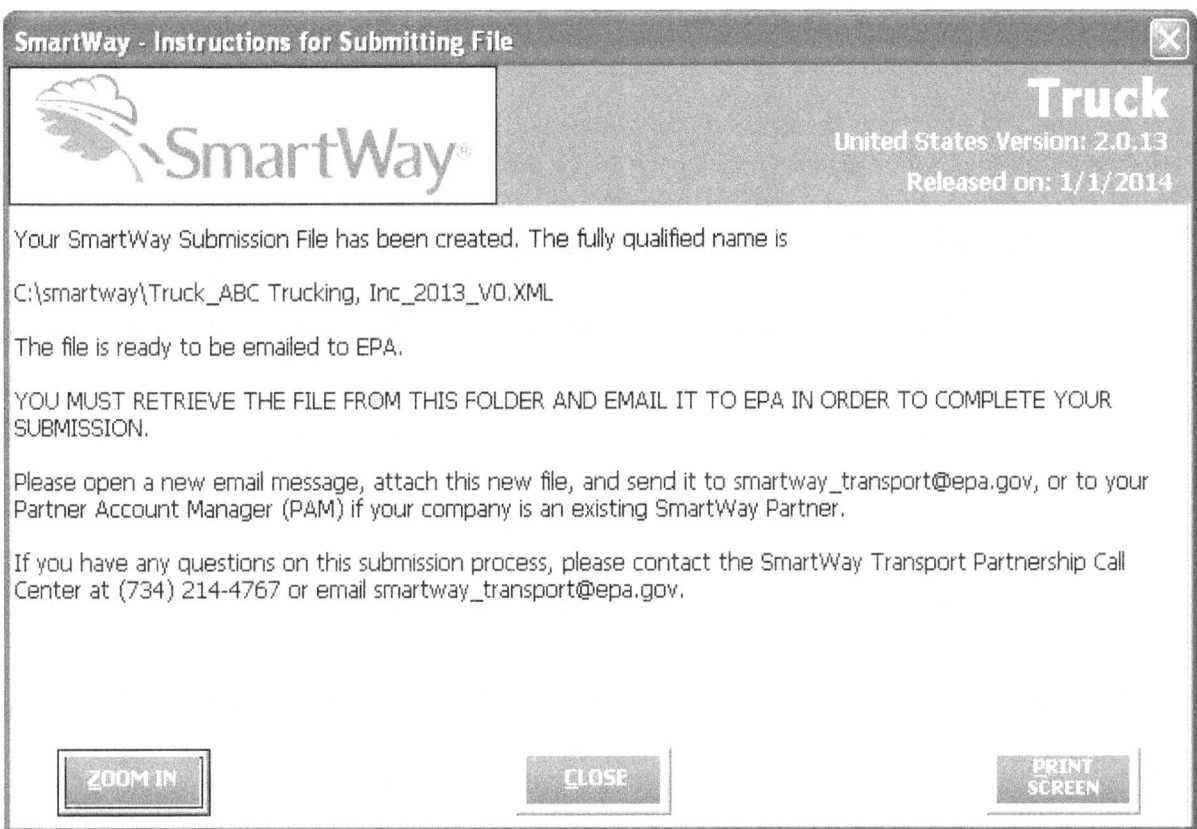

Figure 67: Instructions for Submitting xml File to EPA

Follow these instructions for submitting your .xml file to SmartWay. Note that the .xml file is approximately 10 times smaller than the .xls file itself. Upon selecting NEXT, a screen will appear that allows you to close the Truck Carrier Tool or return to the Home screen.

NOTE: DO NOT ZIP the File. Send it to EPA as a normal file attached in an e-mail. EPA security will not allow zipped files through the EPA firewall.

NOTE: DO NOT CHANGE THE NAME OF THE XML FILE.

NOTE: DO NOT DELETE YOUR EXCEL (XLS) TOOL – SAVE THIS FILE FOR YOUR REFERENCE.

Troubleshooting the Tool

Although the SmartWay tools have been tested extensively, you may encounter errors. Intermittent errors have been encountered when opening the Tool from an e-mail rather than from a drive. In addition, note that you will not be able to open the Tool successfully directly from the SmartWay website, so save the Tool to your hard drive before opening. If you encounter an error during use of the Tool, try restarting the Tool directly from a disk drive, with all other Excel files and applications closed. In addition, make sure that your computer is using a system and application version validated for use with the SmartWay tools (Windows XP or later operating system, and Excel Office 2003 or 2007).

If you continue to encounter problems, make a screen capture of the error message, and save the Tool at that point. (You can make a screen capture by pressing the keys *Alt and Print Screen at the same time*, and then pasting the image into a document such as MS Word.) Then send the screenshot, along with the saved Tool to your Partner Account Manager for further assistance.

Proper Calculation Setting

By default, Microsoft Excel is set to automatically calculate formulas. If values within your Tool are not calculating automatically, then your Excel may be set to "Manual". Follow these instructions to change your Excel Calculation setting.

Office 2010
Choose the "Formulas" tab. Select the "Calculation Options" button in the "Calculation" section to view the drop-down list of options. Choose "Automatic" from the list of options to switch to automatic calculation.

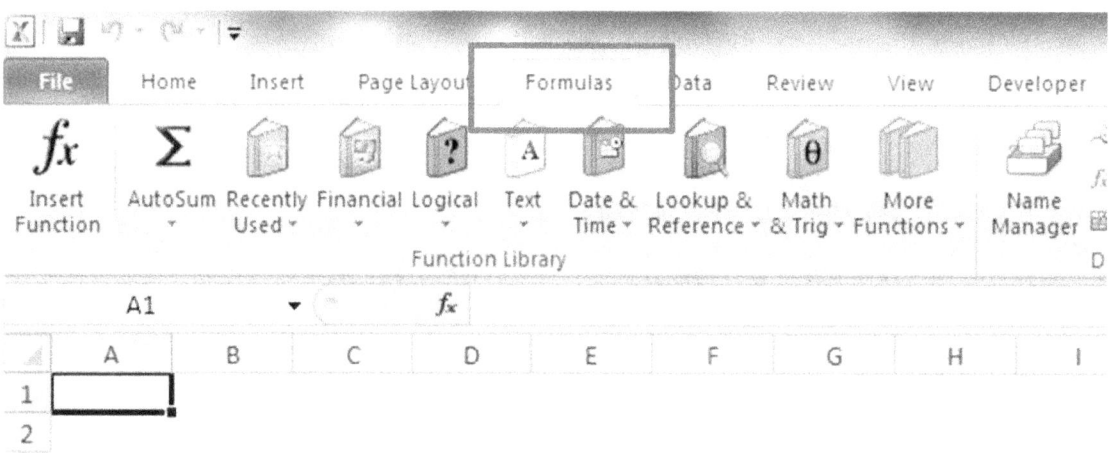

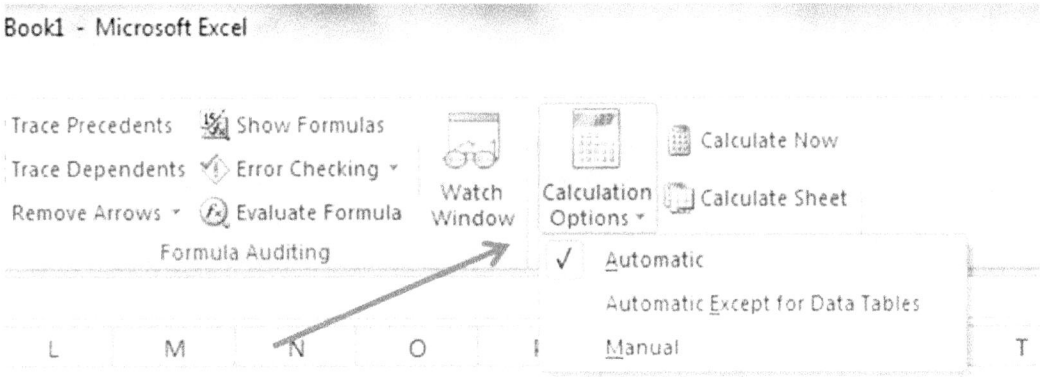

Office 2007

Select the Office button at the top left. Then select the Excel Options button and the Excel Options dialog box will appear. Select the Formulas tab and the Formulas options will appear in the right pane. Select the "Automatic" radio button in the Calculation options section.

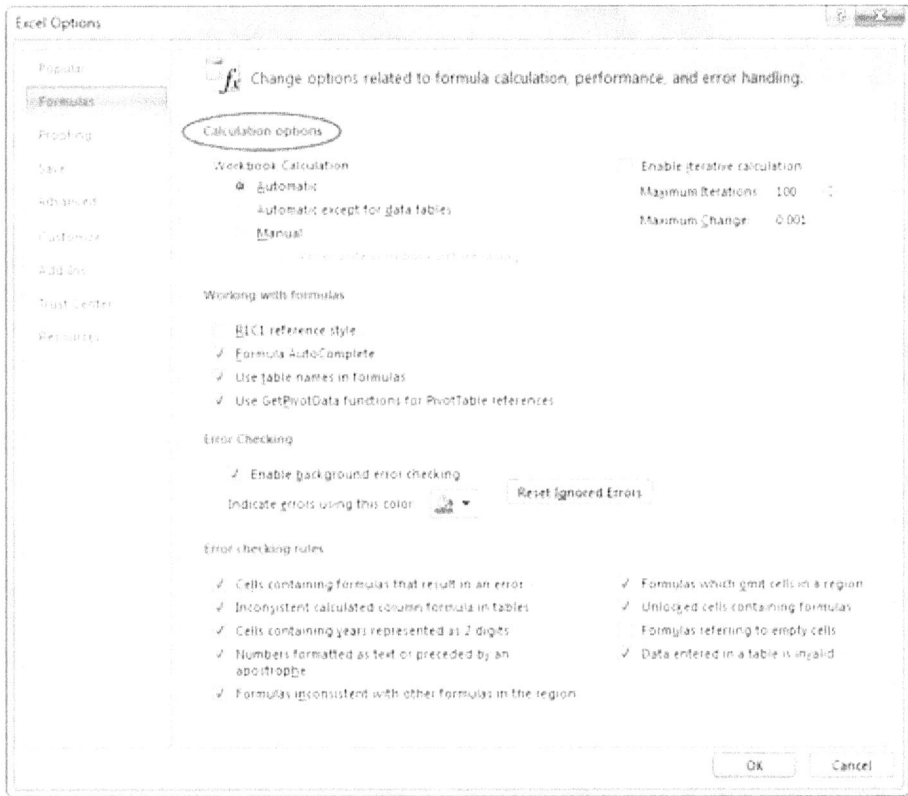

Office 2003

From the tools menu, select "Options". When the Options dialog box is displayed, select the "Calculation" tab. Under the Calculation section, select the "Automatic" radio button and then select "OK".

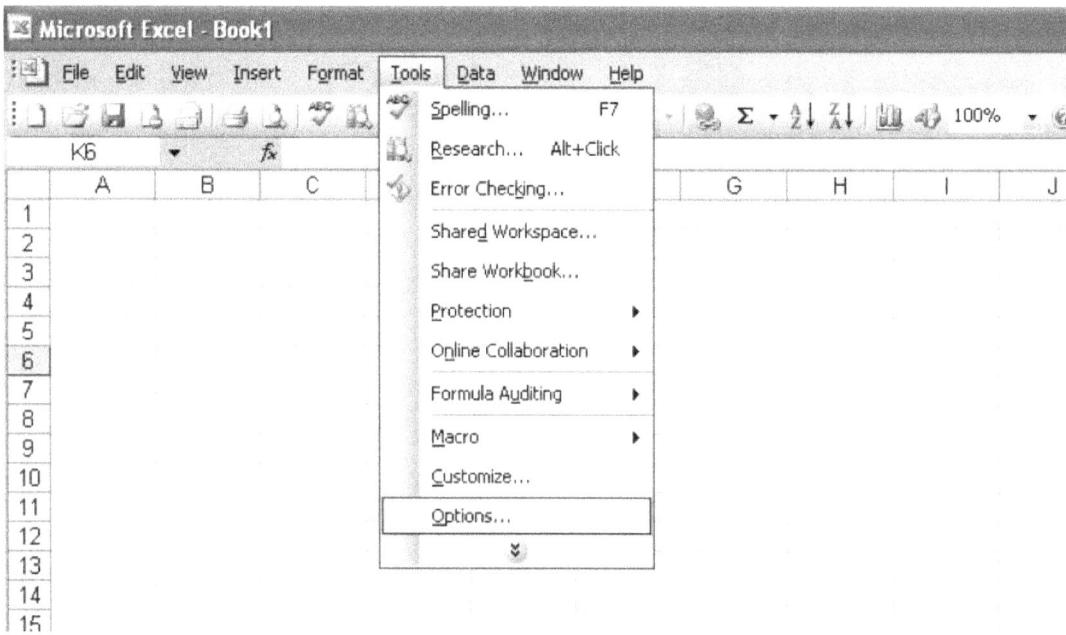

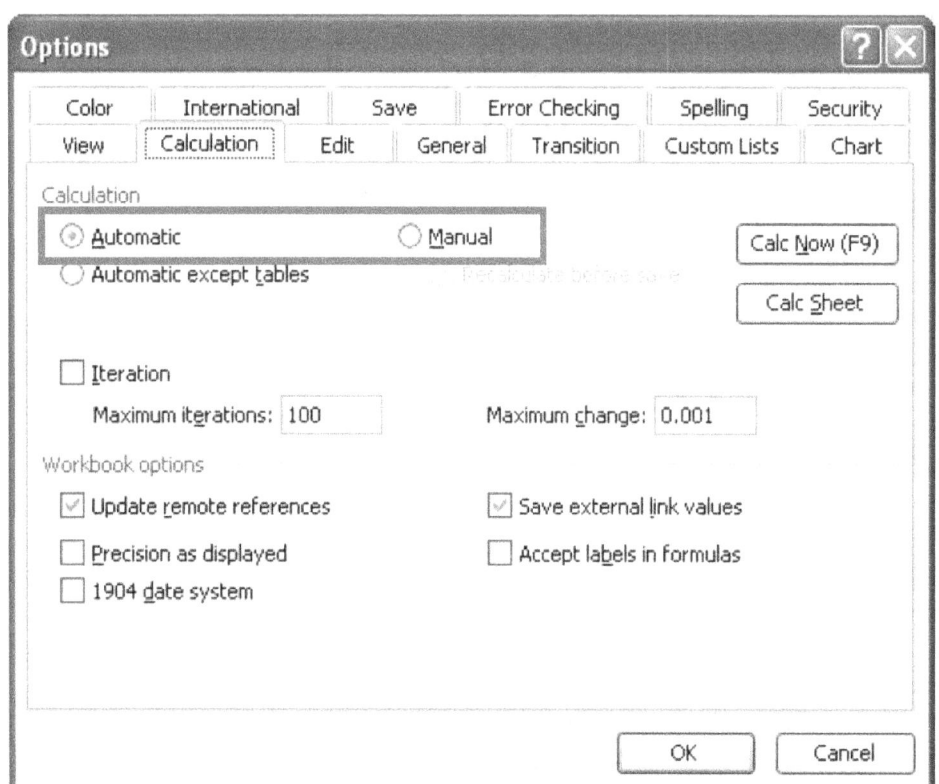

Office 2013 Text Display

Users of MS Office 2013 may encounter problems displaying text entries. The example below shows how the information typed into the "Goal for SmartWay participation" text box is barely visible. This is due to a bug in the MS Office system itself.

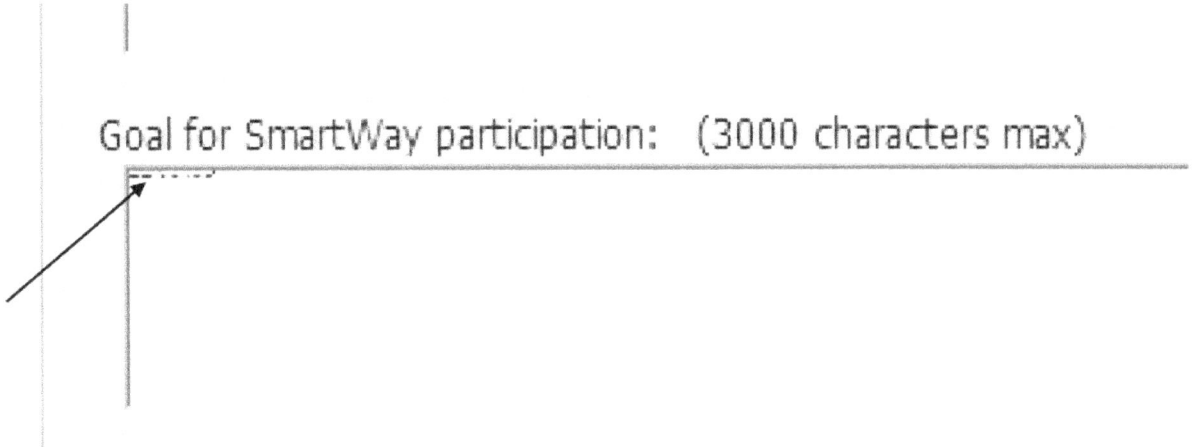

EPA has corrected the display problem in certain locations within the Tool, but if you encounter this problem we recommend opening a different word processor such as MS Word, typing the desired entry in the word processor, then cutting and pasting the entry back into the text box within the Tool. This should allow you to view your entry properly within the Tool.

Appendix A – Port Dray Program Data Requirements and Calculations

EPA's SmartWay Drayage Program recognizes Partners for reducing diesel emissions from port drayage trucks. Drayage is the short distance transport of containerized freight goods between marine (or rail) terminals and local distribution centers. The following provides an overview of the program and the associated data input requirements for the Truck Carrier Tool.

The Partnership Elements: EPA has developed an emissions calculator to help port truck carriers measure their particulate matter (PM), oxides of nitrogen (NOx), and carbon dioxide (CO_2); identify strategies to reduce harmful diesel emissions and track emissions performance on an annual basis. To participate, drayage carrier partners sign a partnership agreement and commit to track emissions, replace older dirtier trucks with cleaner, newer ones, and achieve at least a 50% reduction in PM and 25% reduction in NOx, below the industry average, within three years. Shippers who are already members of SmartWay commit to ship 75% of their port cargo with SmartWay dray carriers within three years. New Shipper Partners must sign the SmartWay Shipper partnership agreement and the drayage addendum.

Business-to-Business Advantage: SmartWay drayage carriers are preferred by SmartWay Shippers, and they want to ship their goods with participating SmartWay drayage partners. Shipper Partners are already giving business priority to SmartWay drayage carriers.

Technical Support: EPA will assist SmartWay drayage partners in developing and meeting goals.

Recognition of Existing Environmental Improvements: A dray carrier fleet's existing age and emissions strategies, coupled with continued improvements, determine status in the Partnership. Full credit is provided for fleet improvements made to date.

Promotional Opportunities and Public Recognition: The SmartWay Transport Partner brand of excellence is awarded to qualifying Partners as a visible cue to business customers and the public, to use in their advertising and other promotional media. The SmartWay brand sends a message that participating dray carriers are champions of environmental stewardship. The SmartWay brand also helps customers and consumers to make educated choices about SmartWay recognized products and services. Visible exposure through national and regional events, advertisements, articles, and special recognition are just a few ways that EPA commits to recognize dray carrier partner achievements.

Truck Carrier Tool Data Input Requirements: Those fleets with 75% or more of their operation in the Dray Operation Category are eligible to participate in SmartWay's Port Drayage Program. These fleets have the **Step 6 – Port Dray Program** option displayed on the General Information screen (see **Figure A-1**).

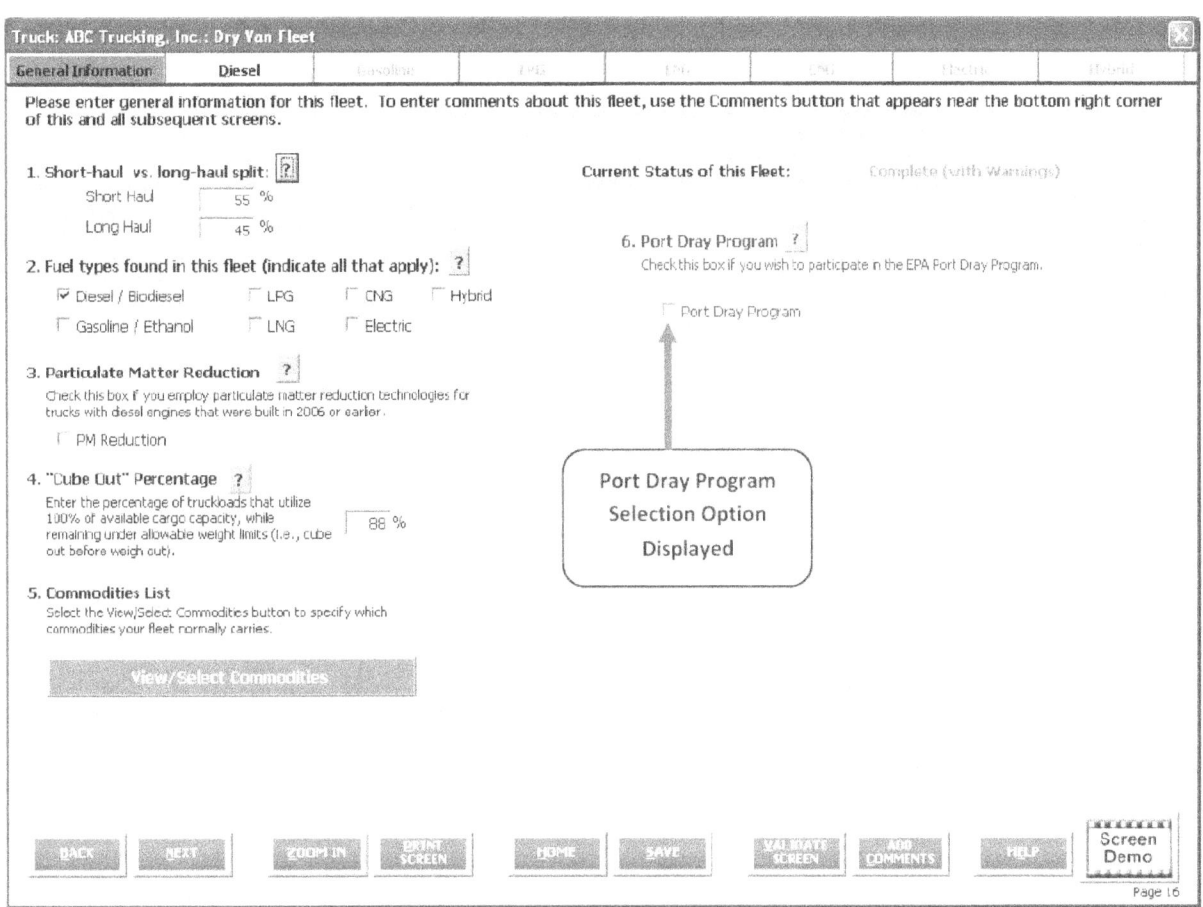

Figure A-1: General Information Screen with Port Dray Program Option

Checking the Port Dray Program box will activate the **Port Dray Program Agreement** screen shown in **Figure A-2**. Select the box on this screen indicating you understand and agree to the terms described therein, and select **OK** to return to the General Information screen.

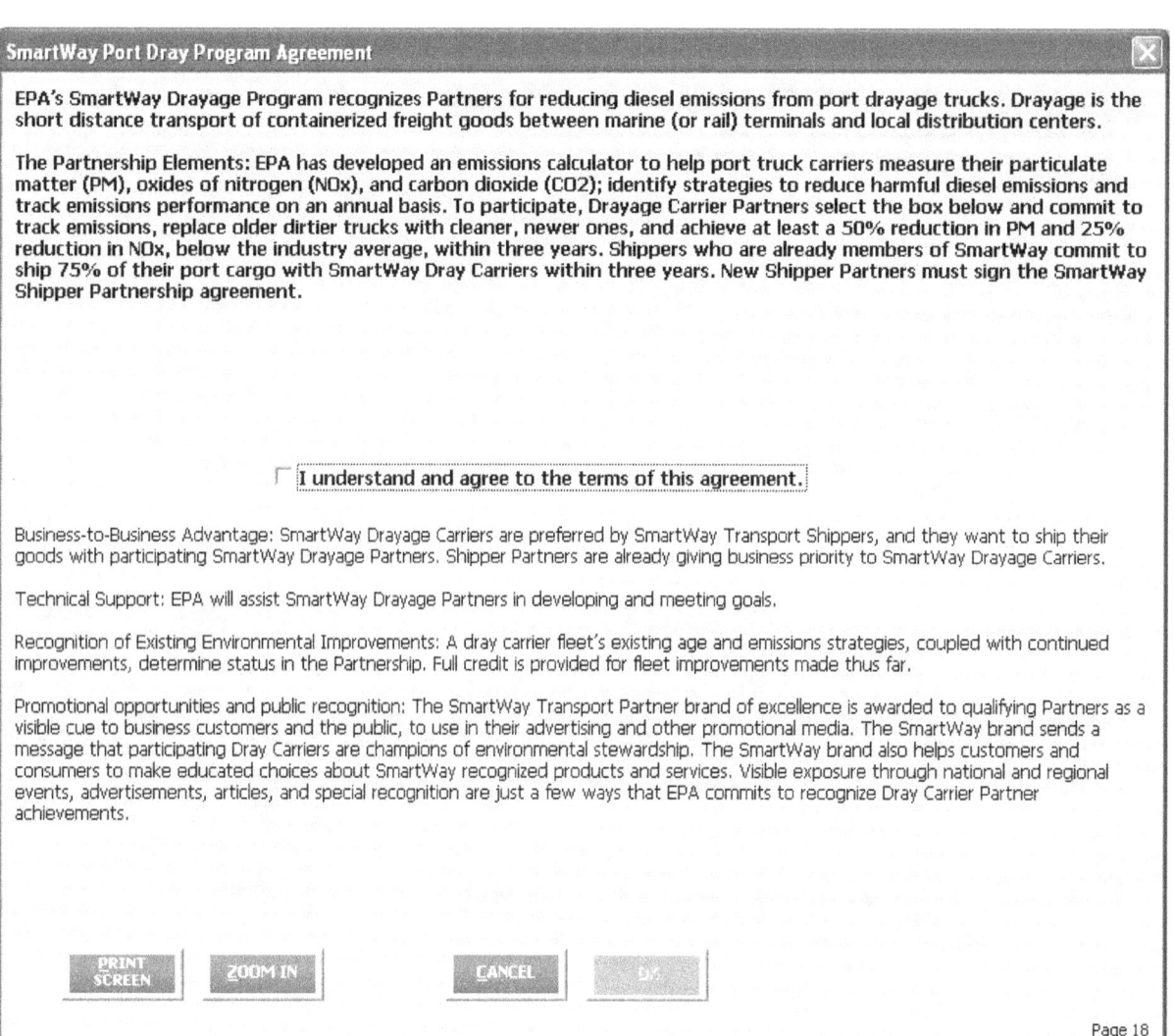

SmartWay Port Dray Program Agreement

EPA's SmartWay Drayage Program recognizes Partners for reducing diesel emissions from port drayage trucks. Drayage is the short distance transport of containerized freight goods between marine (or rail) terminals and local distribution centers.

The Partnership Elements: EPA has developed an emissions calculator to help port truck carriers measure their particulate matter (PM), oxides of nitrogen (NOx), and carbon dioxide (CO2); identify strategies to reduce harmful diesel emissions and track emissions performance on an annual basis. To participate, Drayage Carrier Partners select the box below and commit to track emissions, replace older dirtier trucks with cleaner, newer ones, and achieve at least a 50% reduction in PM and 25% reduction in NOx, below the industry average, within three years. Shippers who are already members of SmartWay commit to ship 75% of their port cargo with SmartWay Dray Carriers within three years. New Shipper Partners must sign the SmartWay Shipper Partnership agreement.

☐ I understand and agree to the terms of this agreement.

Business-to-Business Advantage: SmartWay Drayage Carriers are preferred by SmartWay Transport Shippers, and they want to ship their goods with participating SmartWay Drayage Partners. Shipper Partners are already giving business priority to SmartWay Drayage Carriers.

Technical Support: EPA will assist SmartWay Drayage Partners in developing and meeting goals.

Recognition of Existing Environmental Improvements: A dray carrier fleet's existing age and emissions strategies, coupled with continued improvements, determine status in the Partnership. Full credit is provided for fleet improvements made thus far.

Promotional opportunities and public recognition: The SmartWay Transport Partner brand of excellence is awarded to qualifying Partners as a visible cue to business customers and the public, to use in their advertising and other promotional media. The SmartWay brand sends a message that participating Dray Carriers are champions of environmental stewardship. The SmartWay brand also helps customers and consumers to make educated choices about SmartWay recognized products and services. Visible exposure through national and regional events, advertisements, articles, and special recognition are just a few ways that EPA commits to recognize Dray Carrier Partner achievements.

PRINT SCREEN   ZOOM IN   CANCEL   OK

Page 18

Figure A-2: Port Dray Program Agreement

Upon returning to the General Information screen you will see additional information and data input cells associated with the Port Dray Program (see **Figure A-3**).

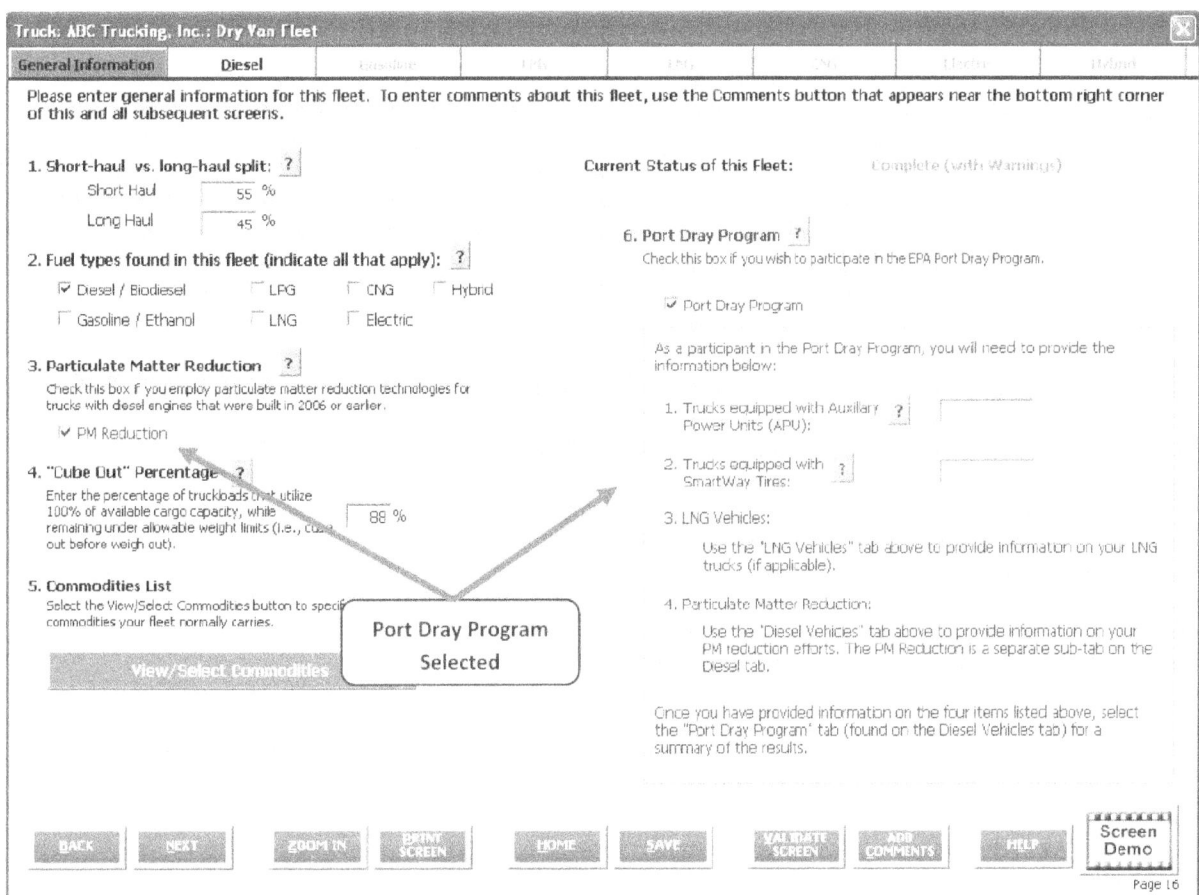

Figure A-3: General Information Screen with Port Dray Program Information and Inputs Activated

In order to participate in the program, first enter the number of auxiliary power units (APUs) installed on your fleet's trucks. An APU is a small diesel fuel-powered generator mounted outside the cab that provides the operator heat, air conditioning, and/or electrical power to run in-cab appliances. Next, input the number of trucks equipped with SmartWay Tires. Options include both dual tires and single wide tires (single wide tires replace the double tire on each end of a drive or trailer axle, in effect turning an 18-wheeler into a 10-wheeler). Low rolling resistance tires can be used with lower-weight aluminum wheels to further improve fuel savings. Enter zero in each of these cells if these strategies are not used on your vehicles. Otherwise, make sure that the number of trucks specified does not exceed the total number entered on the Engine Model Year & Class screen.

If your fleet includes LNG vehicles, make sure to characterize those vehicles on the LNG screen.[13] In addition, note that the **PM Reduction** box on the General Information screen is automatically checked upon selection of the **Port Dray Program** checkbox (see **Figure A-3**). Make sure to characterize your dray fleet's PM reduction strategies on this screen.

---

[13] Port Dray Program participants operating only LNG trucks must still select the Diesel/Biodiesel checkbox in order to view their Port Dray Program score (displayed as a subtab under Diesel).

After entering the required data regarding your dray fleet, the Truck Carrier Tool will calculate your fleet's **Environmental Performance Score** consistent with the Port Dray Program evaluation guidance. See http://www.epa.gov/smartway/partnership/drayage.htm for additional details regarding the program. Information regarding how the **Environmental Performance Score** is calculated is presented in the **Truck Carrier Tool Technical Documentation**. You can view a summary of your Dray Program inputs and the associated **Environmental Performance Score** on the Port Dray Program screen under the **Diesel Vehicles** section, adjacent to the PM Reduction tab (see **Figure A-4**), or by reviewing the Port Dray Program Report on the Reports Menu screen.

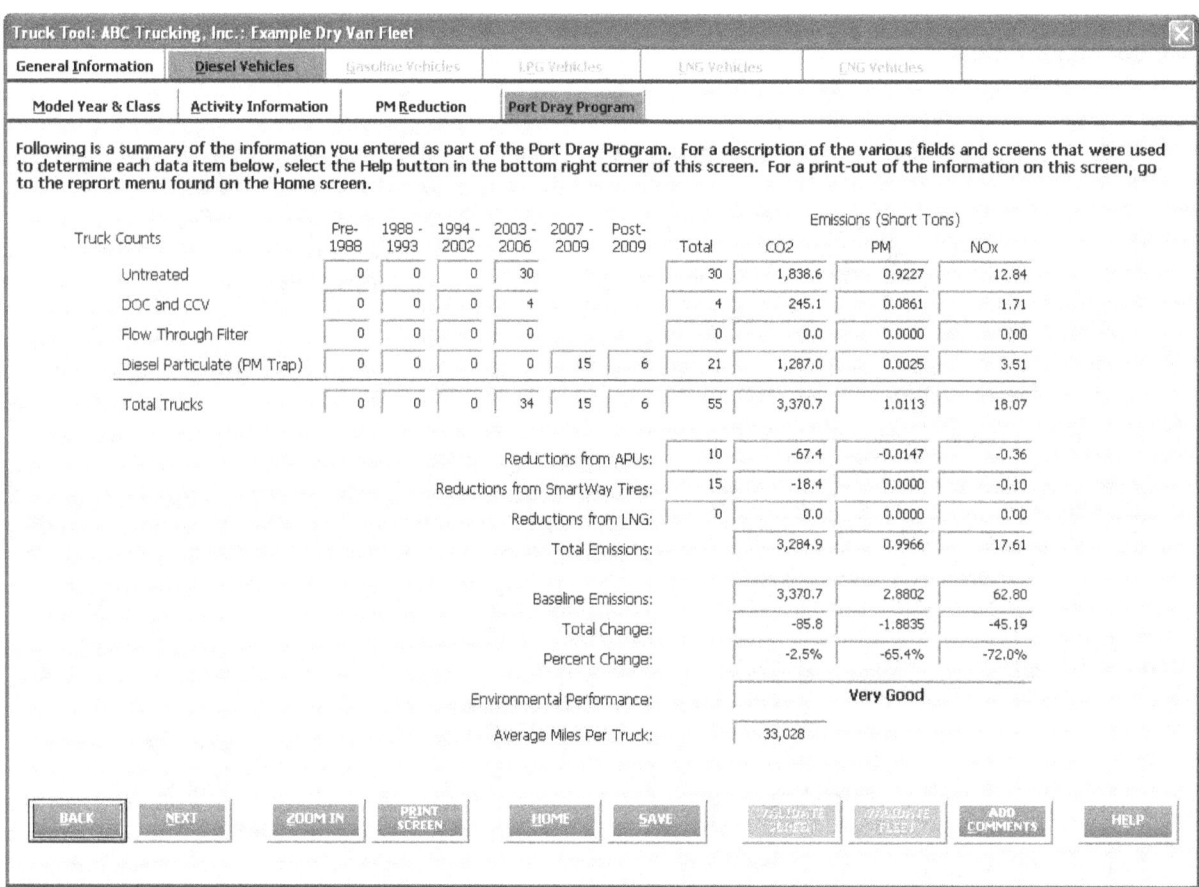

Figure A-4: Port Dray Program Information Screen

The information summarized on the screen above is derived from the following screens within the Truck Carrier Tool. Fields not specifically mentioned below are calculated fields:

1. Average Miles per Truck: Represents "Total Miles Driven" on the Activity Information screen divided by "Total Trucks" on the Engine Model Year & Class screen for diesel vehicles.

2. Average Gallons of Fuel per Truck: Represents "Gallons of Diesel Used" on the Activity Information screen divided by "Total Trucks" on the Engine Model Year & Class screen for diesel vehicles.

3. DOC and CCV: Number of trucks retrofitted is obtained from the PM Reduction screen.

4. Flow Thru Filter: Number of trucks retrofitted is obtained from the PM Reduction screen.

5. Diesel Particulate (PM Trap): Number of trucks retrofitted is obtained from the PM Reduction screen.

6. Untreated: Represents vehicle count totals from the Engine Model Year & Class screen for diesel vehicles, minus truck count totals from the PM Reduction screen (i.e., the number of trucks without retrofits).

7. Reductions from APUs: Number of trucks equipped from the **Port Dray Program** field on the General Information screen.

8. Reductions from SmartWay Tires: Number of trucks equipped from the **Port Dray Program** field on the General Information screen.

9. Reductions from LNG: Represents "Total Trucks" on the Engine Model Year & Class screen for LNG vehicles.

Appendix B—Recommended Data Sources for Activity Data

Table 1 summarizes <u>the standard Data Source categories</u> available for selection for each data type.

Table 1: Data Source Detail Selection Options

<u>Data Type</u>	<u>Data Source</u>	<u>Data Source Detail</u>
Total Miles Driven	As reported to IFTA Form 441 for tax reporting (interstate) – for Class 7, 8a and 8b trucks only	Collected via fleet-wide GPS reporting software
		Collected via odometer readings
		Collected via maintenance records
		Collected via driver trip sheets
		Collected via standard mileage routes, e.g. PC Miler, Household Goods Guide
	As reported to individual states for tax reporting (intrastate)	Collected via fleet-wide GPS reporting software
		Collected via odometer readings
		Collected via maintenance records
		Collected via driver trip sheets
		Collected via standard mileage routes, e.g. PC Miler, Household Goods Guide
	Determined using software	Dispatching Software*
		Transportation Management System (TMS)*
	Vehicle-based data collection	Determined via Electronic Control Module (ECM) data recorder/logger*
Revenue Miles Driven	As used in Federal tax reporting	Collected via electronic Transportation Management System (TMS)
		Collected via GPS-enabled TMS
		Collected via manual input into company database with driver trips sheets
		Collected via non-electronic company records
	Financial data	Accounting/billing software*
		Tax reports/IRS and State*

Data Type	Data Source	Data Source Detail
Revenue Miles Driven (cont'd)	Determined using software	Dispatching Software*
		Transportation Management System (TMS)*
	Based on total mileage	Equal to total miles
		Total miles less empty miles
		Calculated as a percentage of total miles*
Empty Miles Driven	Collected automatically / electronically / manually	Collected via electronic Transportation Management System (TMS)
		Collected via GPS-enabled TMS
		Collected via manual input into company database
		Collected via non-electronic company records
		Collected via odometer readings
		Collected via driver trip sheets
	Financial Data	Accounting/billing software*
		Tax reports/IRS and State*
	Determined using software	Dispatching Software*
		Transportation Management System (TMS)*
	Based on total mileage	Total miles less revenue miles
		Calculated as a percentage of total miles*
Gallons of Fuel Used^	As reported to IFTA Form 441 for tax reporting (interstate) – for Class 7, 8a and 8b trucks only	Collected via electronic fuel receipt
		Collected via paper fuel receipt
		Collected via driver trip sheets
		Collected via electronic expenditure data
		Collected via paper expenditure data
	As reported to individual state for tax reporting (intrastate)	Collected via electronic fuel receipt
		Collected via paper fuel receipt

Data Type	Data Source	Data Source Detail
Gallons of Fuel Used (cont'd)		Collected via driver trip sheets
		Collected via electronic expenditure data
		Collected via paper expenditure data
	Determined using software	Dispatching Software*
		Transportation Management System (TMS)*
	Vehicle-based data collection	Determined via Electronic Control Module (ECM) data recorder/logger*
	Based on MPG estimates*	User-provided
Average Payload	Bills of Lading – electronic records (preferred)	Based on actual miles traveled by specific payloads*
		Trip-weighted (total payload weights / total trips)*
	Bills of Lading – manual records	Based on actual miles traveled by specific payloads*
		Trip-weighted (total payload weights / total trips)*
	Ranges provided by calculator	N/A (calculator)
Average Volume	Determined using company records*	User-provided
	Defaults from calculator	N/A (calculator)
Capacity Utilization	Collected automatically / electronically / manually	Collected through load volume information
	Determined using software	Dispatching Software*
		Transportation Management System (TMS)*
Road Type / Speed Categories	Collected automatically / electronically	Driver trip sheets*
		Governed speed*
		Determined via GPS
		Determined via Electronic Control Module (ECM) data recorder/logger*
	Transportation Management	Driver trip sheets*

Data Type	Data Source	Data Source Detail
	System (TMS)	Governed speed*
		Determined via GPS
		Determined via Electronic Control Module (ECM) data recorder/logger*
Average Annual Idle Hours per Truck	Vehicle-based data collection	Determined via Electronic Control Module (ECM) data recorder/logger*
	Driver trip reports*	User-provided
	Idle reduction strategy	Company "No Idle" policy in place*
		Local/State idle regulation in place*
	Determined using software	Dispatching Software*
		Transportation Management System (TMS)*

* User must provide additional description regarding data collection system and calculation method.

^ For electric vehicles, in lieu of gallons, use kWhrs. Common data sources for kWhrs include metering at charging dock, vehicle data acquisition units, and smart-meter applications.

www.ingramcontent.com/pod-product-compliance
Lightning Source LLC
Chambersburg PA
CBHW081502170526
45166CB00008B/2518